THE CONSTITUENTS OF MEDICINAL PLANTS

An Introduction to the Chemistry

&

Therapeutics of Herbal Medicines

Andrew Pengelly
Medical Herbalist

Sunflower Herbals
2nd Edition

DISCLAIMER

This book is intended for educational and reference purposes, and is not provided in order to diagnose, prescribe or treat any illness or injury. The information contained in the book is technical and is in no way to be considered as a substitute for consultation with a recognised health-care professional. As such the author and others associated with this book accept no responsibility for any claims arising from the use of any remedy or treatment mentioned here.

First published in 1996 by
Sunflower Herbals
Stanley Cottage
Merriwa NSW 2329
Phone (065) 48 5189

2nd Edition, 1997   Reprinted June 1997, March 1998

National Library of Australia Cataloguing-in-Publication entry:

Pengelly, Andrew, 1949- .
The constituents of medicinal plants: an introduction to
the chemistry & therapeutics of herbal medicines.

Bibliography.
ISBN 0 646 28498 3 (1st Edition)
ISBN 0 646 31595 1 (2nd Edition)

1. Herbs - Therapeutic use. 2. Medicinal plants. I. Title.

615.321

Designed and printed by
Fast Books
(a division of Wild and Woolley Pty Ltd)
16 Darghan St, Glebe NSW 2037

Cover photographs: Nick Burgess

CONTENTS

Foreword

The *herbal renaissance* has been taking place throughout the Western world arguably over the last 25 years. Over that period of time herbal medicine has moved from a strong *empirical* basis to its present position as an increasingly *scientifically* based system of healing, increasingly referred to as *phytotherapy*. With the development of the scientific explanation of herbal medicine has come increased credibility and acceptance for it as a system of healing applicable to contemporary society. Andrew Pengelly's book, *THE CONSTITUENTS OF MEDICINAL PLANTS - An Introduction to the Chemistry & Therapeutics of Herbal Medicines* is a significant contribution to the modern day scientific explanation of herbal medicine and as such has made a significant contribution to this increased credibility and acceptance of herbal medicine.

As a long standing graduate of Southern Cross Herbal School, with many years of professional practice as a medical herbalist and as a lecturer on herb identification and introductory pharmacognosy, Mr Pengelly is eminently qualified to write a text such as the above. The book will be particularly suitable to those undertaking professional level studies in herbal medicine or to those in the natural drug industry who seek fundamental information on the phytochemistry of medicinal herbs.

I am delighted and honoured to write this foreword for Mr Pengelly's book and I commend it highly to all those who are interested in understanding the scientific basis of herbal medicine.

Denis Stewart
Medical Herbalist
Director and Senior Lecturer of
Southern Cross Herbal School, Australia

Acknowledgements

Completion of this text book would not have been possible without the contributions and assistance of several people.

I am indebted to Anne Cowper for reading and commenting on the original manuscript, and for the general editorial work carried out. Hans Wohlmuth gave invaluable assistance with the chemical formula diagrams. I also wish to thank the National Herbalist Association of Australia for giving me the use of their excellent library resources and their general support.

By introducing a comprehensive herbal pharmacognosy course the Newcastle College of Herbal Medicine acknowledged the need for a high standard of understanding in this field of education amongst graduate herbalists, in the process giving me the opportunity to write the course material from which this book has developed. For this I would like to thank the principal and school founder Nancy Evelyn.

I am forever indebted to the Southern Cross Herbal School and its director Denis Stewart for my herbal education and for introducing the teaching of pharmacognosy in Australia.

My wife Sunny is a herbal "true believer", and I wish to express my thanks for her constant support and encouragement, without which this book would have been just another idea.

Preface

The main purpose of writing this book was to present a relatively simple introduction to the chemical constituents of herbs for students of medical herbalism in Australia. Since the first edition in 1996, this text has been included in the curriculums of most natural therapy colleges in this country, as well as in New Zealand and at least one U.S. institution. Principles and teachers from several of these colleges have observed the lack of an existing appropriate text on the subject, the only comparable texts being far more comprehensive and costly, usually aimed at the post-graduate scientist or pharmacist rather than the undergraduate. Nevertheless the book has also proven popular with graduates and health-care practitioners, many of whom enjoy having a ready reference to the complex area of plant constituents and therapeutic activities associated with them.

The purpose of this second edition is not to present new material, rather to tidy up a number of typographical and spelling errors that slipped through the proof reading in the first edition. In the nine months since first publication, the text has been carefully scrutinized by various experts and authorities in herbalism and plant chemistry, and apart from some minor disagreements no technical errors have been identified. This is important since I anticipate the book will remain a standard reference in this field for many years to come.

In presenting material of the nature of the topic under discussion, there is always a danger of becoming obsessed with the chemical structures themselves, and presenting these as the main rationale for the therapeutic activities of plant medicines. While not underestimating their influence I believe the constituents are but one explanation of a herb's activities, not necessarily more or less relevant than other factors. I am primarily a holistic practitioner but I do not believe we can disregard the increasing information becoming available to us through modern research, even though this research may be based on reductionist principles. It is encumbent on practitioners and teachers of plant medicine to stay abreast of the research, but our interpretation and application of the research should be based on a holistic understanding of the herbs and of disease processes.

Chapter 1
INTRODUCTION TO PHYTOCHEMISTRY

PHYTOCHEMICAL BASIS OF HERBAL MEDICINES

There are two fundamental ways in which herbal medicines may be classified – according to their energetic qualities and their phytochemical constituents. Energetic qualities are highly valued in traditional systems such as Ayurvedic and Traditional Chinese Medicine, but less so in Western medicine. Through application of pharmacognosy – involving study of herbal medicines from all perspectives – it is possible to merge the scientific with the energetic. For the equivalent in Western herbalism ie. merging of Eastern and Western use of herbs, readers should consult *The Energetics of Western Herbs* by Peter Holmes and Michael Tierra's *Planetary Herbology*.

To the scientist or pharmacist a plant's constituents may compose of, to use the words of Simon Mills "a crude chemical brew", so the search is always on for an identifiable "active principle" [Mills 1994]. Herbalists on the other hand aim at a holistic approach which values the sum of the constituents – even those considered by pharmacists to be worthless. Many of the studies referred to in this book are based on this reductionist approach to research, however this does not necessarily devalue the results of this research, since isolation and experimentation with single constituents can provide information that can be adapted to a more holistic understanding of a herbs action. Knowledge of individual constituents is also essential for developing quality control methods, extraction procedures, understanding of pharmacological activity and pharmacokinetics [Bruneton 1995]. It is not merely a necessary step in the isolation and synthesis of plant derived drugs.

UNDERSTANDING ORGANIC CHEMISTRY

It does not require a science degree to gain an understanding of the fundamental chemical structures found in medicinal herbs, but some knowledge of organic chemistry is desirable. The most user-friendly text for explaining the chemistry of herbs is Terry Willard's *Textbook of Advanced Herbology*, while Varro Tyler's *Pharmacognosy* is an essential reference. Regularly browsing through peer review journals such as *Phytochemistry* and *Tetrahedon Letters* (available in university libraries) also provides good educational value.

Primary and secondary metabolites

Primary plant compounds are of universal occurrence ie. they are found in all or most plant species. They are products of vital metabolic pathways especially the citric acid/ krebs cycle and include proteins, fats, sugars, organic acids. The energy released by glycolysis of carbohydrates and through the citric acid cycle fuels the synthesis of all organic plant compounds or phytochemicals, including the secondary metabolites [Samuelsson 1992].

Secondary compounds were once regarded as simple waste products of a plants metabolism. However this argument is weakened by knowledge of the existence of specialist enzymes, strict genetic controls and the high metabolic requirements of these compounds [Waterman & Mole 1994]. Today most scientists accept that many of these compounds serve primarily to repel grazing animals or destructive pathogens [Cronquist 1988].

The main metabolic pathways through which the chemical compounds are synthesized in the body are more or less universal in the plant kingdom. Metabolic pathways involve series of enzymes which are specific for each compound.

shikimic acid

The most common pathways are:
- shikimic acid – produces phenols, tannins, aromatic alkaloids
- acetate-malonate – phenols, alkaloids
- mevalonic acid – terpenes, steroids, alkaloids

Isomerism

The mystery surrounding organic chemical structures is partly due to the three dimensional shapes of these molecules, allowing for two or more positions of atoms on the same basic molecule. Stereoisomers are simply compounds whose side chains are attached at different locations around the carbon ring. For example the phenol **coumaric acid** may contain a hydroxyl (OH) group at any of three locations, known as *ortho* (0-coumaric acid), *meta* (*m*-coumaric acid) or *para* (*p*-coumaric acid).

Optical isomers are chemicals such as **lactic acid** which have different varieties whose only difference is their effect on polarized light. When polarized light is passed through a molecule of lactic acid the plane of polarization may rotate to the right (dextrorotary: *d* or +lactose) or the left (laevorotary: *l* or -lactose). However despite this apparent minor difference the two isomers produce different reactions in the human body.

An atom or group attached to the ring below the plane is termed alpha (α) and is usually denoted with broken lines. When attached above the plane it is called beta (β) and is shown as a heavy solid line.

ORGANIC ACIDS

Organic acids are of such widespread occurrence they are not strictly secondary metabolites, in fact many occur during the citric acid or Krebs cycle. They are water soluble colourless liquids with characteristic sharp tastes.

Monobasic acids contain a single carboxyl group (-COOH), they include the fatty acids as well as **isovaleric acid**, a sedative principle found in *Valeriana officinalis* and *Humulus lupulus*. One of the most important in this group is **acetic acid**, the main constituent in vinegar. Acetic acid is the precursor of lipids as well as some essential oils and alkaloids.

Polybasic acids contain two or more carboxyl groups, they generally have a slight laxative effect. They include **oxalic acid** and **fumaric acid**, the latter occurring in *Fumaria officinalis*.

Hydroxyl acids include a hydroxyl group (OH) with a pair of carboxyl groups. **Citric** and **tartaric acids** are the most common examples.

Aromatic acids such as **benzoic acid** are sometimes classed with the organic acids, however they are products of the shikimic acid pathway and are discussed further in the phenols section of this book.

Acyclic, cyclic and heterocyclic compounds

The atoms of organic compounds are arranged either as open chains (acyclic or aliphatic) or as closed ring systems (cyclic). Each corner or kink in the ring (or chain) indicates a CH_2 group, though these are usually abbreviated to C or omitted. Unsaturated rings systems are those in which carbons are linked by double or triple bonds, while saturated rings do not contain any double bonds. The benzene ring, the structure which forms the basis of thousands of organic compounds, is an unsaturated six carbon ring which is generally illustrated as a hexagon containing three double lines. Compounds containing one or more benzene rings are known as aromatic compounds.

Ring systems in which the rings are composed entirely of hydrocarbons are called homocyclic (eg. benzene). Ring systems constituted of different elements are called heterocyclic compounds. Such ring systems usually

Furan

Pyran

Pyrrole

Thiophene

Pyridine

contain several carbon atoms and one or more atoms of other elements, usually nitrogen, oxygen or sulphur. Over 4000 heterocyclic systems are known from plant and animal sources. They sometimes occur fused to a benzene ring or to another heterocyclic ring, to give bicyclic systems. Some of these heterocyclic rings resist opening and remain intact throughout vigerous reactions, as does the benzene ring.

Important parent heterocyclic compounds:

> **Furan**
>
> **Pyran**
>
> **Pyrrole**
>
> **Thiophene**
>
> **Pyridine**

Note that in **furan** and **pyran** the C is replaced by O, while S substitutes for C in **thiophene**.

Pyrrole and **pyridine**, in which C is replaced by N, are the precursors for many of the medically important alkaloids.

Pyrrole ring occurs in many important natural compounds: **chlorophyll** – the central magnesium atom is surrounded by four pyrrole rings connected by four carbon atoms; **haemoglobin**, in which the central iron atom is surrounded by the four pyrrole rings; plant alkaloids including **nicotine** and **cocaine**.

Pyridine and its derivatives constitute the most widely occuring heterocyclic six-atom ring compounds occurring in nature. Many such compounds are also derived from **piperidine**, the reduction product of pyridine. Pyridine occurs naturally in coal tar as well as the following vitamins and alkaloids: vitamin B6 – **pyridoxine**; nicotinic acid – **niacin; piperidine, quinoline** and **isoquinoline** plant alkaloids.

Functional Groups

Another feature of organic compounds is the presence of functional groups. These are groups of atoms attached to carbon chains or rings – they are often involved in chemical reactions. The different classes of functional groups are distinguished by the number of hydrogen atoms replaced. They are named as follows [Karlson 1969]:

- OH (hydroxyl group)   - NH2 (amino group)

= O (carbonyl group)   = NH (imino group)

= O (carboxyl group)   = NH3 (nitrile)

\ OH

References

Bruneton, J. 1995. *Pharmacognosy Phytochemistry Medicinal Plants.* Lavoisier Pubs, Paris.

Cronquist, A. 1988. *The evolution and classification of flowering plants. 2nd ed.* New York Botanical Gardens.

Holmes, P. 1989. *The Energetics of Western Herbs* (2 volumes) Artemisia Press, Boulder, USA.

Karlson, P. 1969. *Introduction to modern biochemistry.* Academic Press, New York.

Mills, S. 1994. *The essential book of herbal medicine.* Viking Pubs, UK.

Nathan, S. & Murthy, M. 1968. *Organic chemistry made simple.* W.H. Allen, London.

Samuelsson, G. 1992. *Drugs of Natural Origin.* Swedish Pharmaceutical Press.

Sharp, D. 1990. *Dictionary of Chemistry.* Penguin books, London.

Tierra, M. 1988. *Planetary Herbology.* Lotus books, Santa Fe, USA.

Tyler, V., Brady, J. & Robbers, J. 1988. *Pharmacognosy 9th edition.* Lea & Febiger, Philadelphia.

Waterman, P. & Mole, S. 1994. *Analysis of Phenolic Plant Metabolites.* Blackwell Scientific Pubs, UK.

Willard, T. 1992. *Textbook of Advanced Herbology.* Wild Rose College of Natural Healing, Alberta.

Chapter 2
PHENOLS & TANNINS

SIMPLE PHENOLS

phenol

pyrogallol

gallic acid

Phenols are one of the largest groups of secondary plant constituents. They are defined as compounds that bear at least one hydroxyl group attached to an aromatic or benzene ring system. In addition the ring system may bear other substitutes especially methyl groups.

Simple phenols consist of an aromatic ring in which a hydrogen is replaced by a hydroxyl group. Their distribution is widespread amongst all classes of plants. They are tertiary alcohols which act like weak acids. General properties of simple phenols are bactericidal, antiseptic and anthelmintic. Phenol itself is a standard for other antimicrobial agents

The simplest phenols are C_6 structures consisting of an aromatic ring with hydroxyl groups attached. These include **pyrogallol** and **hydroquinone**.

Addition of a carboxyl group to the basic phenol structure produces a group of C_6C_1 compounds, including some of widespread distribution amongst plants and important therapeutic activity. The most important of these are **gallic acid** and **salicylic acid**.

A rarer type of simple phenols with C_6C_2 structures are known as **acetophenones**. Some of these have demonstrated antiasthmatic activity, particularly **apocynin** and its glycoside **androsin** which are derived from *Picrorrhiza kurroa* [Dorsch et al. 1994].

PHENYLPROPANOIDS

These are C_6C_3 compounds, made up of a benzene ring with a three carbon side chain. The most important are the hydroxy cinnamic acids: **caffeic acid, *p*-coumaric acid, ferulic and sinapic acids**. They can be derived from different stages of the shikimic acid pathway. These acids are of much benefit therapeutically and are non-toxic. They may also occur as glycosides.

Caffeic acid is an inhibitor of the enzymes DOPA-decarboxylase and 5-lipoxygenase. It is analgesic and anti-inflammatory, and promotes intestinal motility [Adzet & Camarasa 1988]. It is of widespread occurrence and is found in green and roasted coffee beans.

caffeic acid

Cynarin (1.5 dicaffeoyl-D-quinic acid), the major active principle of globe artichoke – *Cynara scolymus* (Asteraceae), is formed from the bonding of two phenolic acids, caffeic and quinic acids. **Cynarin** is a proven hepatoprotective and hypocholesterolaemic agent.

curcumin

Curcumin, the yellow pigment from turmeric rhizome *Curcuma longa* (Zingiberaceae). Curcumin and its derivatives are diarylheptanoids. They have significant anti-inflammatory, hypotensive and hepatoprotective properties [Ammon & Wahl 1990].

hydroquinone

Decarboxylation of caffeic acids results in formation of simple phenols – eg. hydroxybenzoic acid →**hydroquinone**. Upon glucolysation **arbutin**, a simple phenol glycoside, is formed.

Arbutin occurs in leaves of the pear tree (*Pyrus communis*) and *Arctostaphylos uva-ursi*, a urinary tract antiseptic and diuretic. Arbutin is hydrolised to hydroquinone in alkaline urine – this effect is strictly localised.

arbutin

It is indicated for urinary tract infections ie. cystitis, urethritis, prostatitis.

SALICYLATES & SALICINS

Salicylic acid is classed with the simple phenols.

It is a carboxylated phenol ie. carboxylic acid + a hydroxyl group added to a benzene ring.

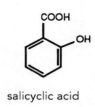

salicyclic acid

Salicylic acid is rarely found freely in plants, but usually occurs as glycosides (salicins), esters and salts. These derivatives are converted to salicylic acid in the human body.

Salicylic acid was first prepared in pure form from meadowsweet – *Filipendula ulmaria* (Family Rosaceae) in 1838. It was first synthesised by Kolbe, a German chemist, in 1860. The subsequent synthesis of acetylsalicylic acid was performed in 1899 by the Bayer company who gave us aspirin.

Main derivatives of salicylic acid
* **Glycosides**
  Salicin – found in willow bark (*Salix* spp.), poplar bark (*Populus* spp.), and *Viburnum* spp.
  Populin – *Populus* spp.
  Gaultherin – Wintergreen (*Gaultheria* spp.)
  Spiraein – *Filipendula* spp.
* **Esters**
  Methyl salicylate – found in meadowsweet (*Filipendula* spp.), *Gaultheria* spp.
  Salicylaldehyde – *Filipendula* spp.
* **Acetylsalicylic acid** (Aspirin) – ASA
  Aspirin is a synthetic derivative of salicylic acid

Properties of salicins and salicylates
Analgesic – long history of use for relief of headaches in European and North American folk medicine.

Dual mode of action (proposed)

- depressant on central nervous system
- influence on prostaglandin metabolism

Antipyretic – act to increase peripheral blood flow and sweat production, by direct action on the thermogenic section in the hypothalamus. Used for neuralgias, sciatica, myalgia, headaches.

Anti-inflammatory – used in rheumatic conditions.

Anti-clotting effect – present in ASA (Aspirin) but not demonstrated with salicins. Through its acetyl group ASA blocks cyclooxygenase in the blood platelets, thus irreversibly inhibiting thromboxane synthesis [Meier & Leibi 1990]. By this rationale salicin containing herbs should not be used as anti-clotting agents in the way Aspirin is presently used in western medicine.

Toxicity

A well known tendency to gastric haemorrhage is associated with use of Aspirin. However there is no evidence of this effect in salicin containing herbs.

nordihydroguaiaretic acid

LIGNANS

These are dimeric compounds in which phenylpropane (C_6C_3) units are linked to form 3-dimensional networks. They are distinct from lignin which is a high molecular weight polymer based on the C6C3 compound coniferyl alcohol.

Lignans are another widespread class of compounds that demonstrate significant therapeutic benefits to humans. The simple lignan with the long name **nordihydroguaiaretic acid (NDGA)** from chaparral – *Larrea* spp. (Zygophyllaceae) is a potent antioxidant.

schizandrin

Shizandrins from *Schizandra chinensis* (Magnoliaceae) reverse destruction of liver cells by inducement of cytochrome P-450 [Huang 1993] while other lignans are antiviral and antineoplastic [MacRae et al 1989; San Feliciano et al 1993].

Podophyllotoxin, a lignan derived from the resin of the mayapple *Podophyllum peltatum* (Berberidaceae), is an anti-mitotic (inhibits cell divisions) and a caustic used topically for warts and papillomas, though extreme care is required due to the highly irritant nature of the compound. The anticancer drugs etoposide and teniposide were synthesised from podophyllotoxin [Hamburger & Hostettmann 1991].

Flavonolignans from *Silybum marianum* are known as hybrid lignans since they are also classed amongst the flavonoids. The mixture of flavonolignans collectively known as silymarin which are found in the fruits of this thistle, have well documented hepatoprotective actions.

Lignans are also present in grains and pulses and their regular consumption has been shown to affect oestrogen levels in humans [Beckham 1995].

COUMARINS

Coumarin is the lactone of O-hydroxy-cinnamic acid, it has a cyclized C_6C_3 skeleton.

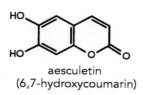

aesculetin
(6,7-hydroxycoumarin)

It occurs as colourless, prismatic crystals and has a characteristic fragrant odour and a bitter, aromatic burning taste. It is soluble in alcohol and can be readily synthesised in the laboratory.

Classification and structures
Most simple coumarins are substituted with OH or OCH_3 at positions C-6 and C-7.

umbelliferone
(7-hydroxycoumarin)

They often occur in glycosidic form eg. **aesculin** is the glycoside of **aesculetin**.

Furanocoumarins have a furan ring at C-6 and C-7 (**psoralen**) or C-7 and C-8 (**angelican**) of the coumarin ring system, however they are not phenolic in structure. The linear furanocoumarins **psoralen** and **bergapten** have photosensitising properties which have been utilised in treatments of vitiligo and psoriasis where the subject is concurrently exposed to solar radiation. These coumarins are found in *Ammi majus* and *Angelica archangelica* (Apiaceae), *Ruta graveolens* and *Citrus* spp. (Rutaceae) as well as the common fig – *Ficus carica* (Moraceae) [Towers 1980].

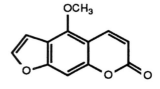

bergapten

The furanochromone **khellin** is the active constituent of *Ammi visnaga* (Apiaceae), a significant antispasmodic and antiasthmatic herb, which also has a beneficial action on coronory blood vessels. Pyranocoumarins, which contain a pyran ring fused at C-7 and C-8, are also present in *Ammi visnaga* [Greinwald and Stobernack 1990].

The distribution of coumarins is widespread. Originally isolated from tonka beans, they are abundant in particular plant families eg. Rubiaceae – *Asperula*; Poaceae – *Avena*; Fabaceae – *Medicago, Melilotus*; Rutaceae – *Ruta, Murraya*; Apiaceae – *Angelica, Ammi*.

Dicoumarol (bishydroxycoumarin), used as a medical drug, was originally derived from *Melilotus officinalis*. **Warfarin**, the blood-thinning drug also used as rat poison, is a synthetic derivative.

Properties of coumarins
Anti-coagulant, antimicrobial, fungicidal.

Antispasmodic – **visnadin, khellin** from *Ammi visnaga* – Calcium channel blockers

Antifertility agents – **chalepensin** from *Ruta graveolens* (Kong et al 1989)

Photosensitising – **psoralen** from *Ammi majus*.

QUINONES

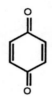

p-quinone

Quinone itself is benzoquinone ($C_6H_4O_2$), a diketone, which undergoes reversible oxidation-reduction (redox) reactions in the presence of reductase enzymes. The reduced form of quinone is **hydroquinone** which, as noted above, occurs as the glycoside **arbutin**. Many of the more complex compounds including some napthoquinones and anthraquinones have phenolic structures.

Quinones form an important component of the electron-transport system in plants and mammals. **Ubiquinol**, the reduced form of co-enzyme Q_{10} and **menaquinone** or vitamin K have significant anti-oxidant properties, playing a major role in protecting cells from free-radical damage [Cadenas & Hochstein 1992]. These two compounds are virtually universal and hence not classed among the secondary metabolites [Bruneton 1995]. Any of four different metabolic pathways may be involved in quinone biosynthesis [Harborne & Baxter 1993].

Many quinones are known to induce contact dermatitis and respiratory reactions in susceptible people. Napthoquinones including **lapachol** and others found in the ebony family (Ebenaceae) have been associated with these reactions [Bruneton 1995].

Napthoquinones

These are dark yellow pigments with interesting pharmacological properties. The hair dye henna is derived from the plant *Lawsonia inermis* (Lythraceae) which contains the napthoquinone **lawsone** linked to a sugar (ie. a glycoside). Other napthoquinones have antimicrobial and antifungal properties. These include **juglone** from the walnut – *Juglans regia* and butternut – *J. cineria* (Juglandaceae), which occurs in leaves and stain derived from the fresh plants [Bruneton 1995]. The leaves are also rich in hydrolysable tannins, hence they are of benefit for piles and venous insufficiency. **Juglone** is a laxative and vermifuge agent.

Other napthoquinones with potent antimicrobial properties are **plumbagin** from the sundew *Drosera rotundifolia* (Droseraceae) and **lapachol** from pau d'arco – *Tabebuia impetiginosa* and spp. (Bignoniaceae).

plumbagin

TANNINS

Tannins represent the largest group of polyphenols. They are widely distributed in the bark of trees, insect galls, leaves, stems and fruit. Tannins were originally isolated from the bark and insect galls of oak trees – *Quercus* spp. (Fagaceae).

Tannins are non-crystalline compounds which in water produce a mild acid reaction. Their ingestion give rise to a puckering, astringent sensation in the mouth. The taste is sour. They often occur as glycosides. Their ability to precipitate proteins into insoluble complexes enables humans to 'tan' animal hides and convert them to leather. It is also the basis of their astringent effects. Due to protein precipitation, the tannins exert an inhibitory effect on many enzymes, hence contributing an antipathogenic protective function in bark and heartwoods of woody plant species [Scalbert 1991]. Tannins also forms precipitates with polysaccharides and some alkaloids including caffeine.

ellagic acid

Tannins are high molecular weight compounds MW500-5000 containing sufficient phenolic hydroxyl groups to permit the formation of stable cross links with proteins, and as a result of this cross-linking enzymes may be inhibited. Almost all tannins are classified as either hydrolysable tannins or condensed tannins. Some plants contain both kinds eg. *Quercus robur*.

Hydrolysable tannins
These are derived from simple phenolic acids, particularly **gallic acid**, which is linked to a sugar by esterfication

[Waterman & Mole 1994]. The gallic acid groups are usually bonded to form dimers such as **ellagic acid**.

Hydrolysable tannins break down on hydrolysis to give gallic acid and glucose (= gallotannins) or ellagic acid and glucose (ellagitannins). They are readily soluble in water and alcohol. Botanicals containing hydrolysable tannins include *Geranium maculatum* (Geraniaceae), *Agrimonia eupatoria* (Rosaceae) and *Arctostaphylos uva-ursi* (Ericaceae).

flavan-3-ols-catechin

Condensed tannins

Condensed or phlobotannins are polymers of flavan-3-ols (catechins) and flavan-3,4-diols (leuco anthocyanins).

Upon hydrolysis condensed tannins form phlobaphenes, insoluble red residues also known as "tanners red". They are only partially soluble in water and alcohol, the addition of glycerine aids solubility.

Botanicals with condensed tannins include *Eucalyptus* spp, *Hamamelis virginica*, *Cinchona* spp., *Acacia* spp., and tea (*Camellia sinensis*).

Astringent action

Astringents cause contraction of tissue, blanching and wrinkling of mucous membranes, and diminished exudations. When applied to wounds they form a thin protective surface, cutting down on secretion of exudates. Precipitation of proteins and/or polysaccharides on the surface results in hardening of the epidermis, reducing absorption of toxins, and protecting against irritants. Tannins display antimicrobial properties. Many are able to constrict blood vessels, thereby reducing bleeding.

Uses of tannins

1. Protecting inflamed mucous membranes
2. Drying effect on mucous membranes, reduces hypersecretions

3. Reduces inflammation and swelling which accompany infections
4. Prevents bleeding from small wounds
5. Reduce uterine bleeding eg. menorrhagia, metrorrhagia
6. Binding effect in the gut – relieves diarrhoea, dysentry.
7. Used externally as douches, snuffs, eyewash.

Some recent findings in research on tannins

1. Anti-viral activity against *Herpes simplex* in vitro. Anti-HIV activity [Okuda et al. 1989].

2. Antimicrobial. Catechin tannin extracts exhibited antibacterial activity against seven tested microorganisms (include. E.coli, Staph. aureus, Salmonella). The activity was lower than that of the reference drug tetracycline [Lutete et al. 1994]. Tannins also inhibit the growth of many filamentous fungi, though yeasts appear to be more resistant [Scalbert 1991].

3. Antioxidant properties. Inhibition of lipid peroxidation found to be strongest in ellagitannins followed by condensed tannins. Inhibits auto-oxidation of ascorbic acid (eg. geranium ellagitannin) [Okuda et al. 1988]. The proanthocyanidins or **pycnogenols**, condensed tannins derived from pine bark and grape seeds are potent anti-oxidants for the vascular system. Similar compounds are found in the hawthorne (*Crataegus* spp.).

4. Inhibition of gastric secretions – ellagic acid reduces occurrence of stress-induced gastric lesions, especially duodenal ulcers [Murakami et al. 1991].

5. Hepatoprotective – 46 tannin compounds from Chinese herbs were assessed employing experimentally induced cytotoxicity in primary cultured hepatocytes. Most tannins exhibited prominent effects in the CCl_4 induced toxicity. Of these the hydrolysable tannins were the most potent [Hikino 1985].

Phenolic compounds in foods

Anthocyanin pigments (condensed tannins) are found in red/blue/black fruits. *Ribes, Rubus & Vaccinium* fruits were shown to possess superoxide radical scavenging and antilipoperoxidant activities [Constantito et al. 1994]. Grape seeds are a major source of the condensed tannins known as oligomeric proanthocyanidins (OPCs) [Bombardelli & Morazzoni 1995].

Condensed tannins are also found in teas and red wine. Green tea is especially rich in the beneficial epigallocatechins [Wohlmuth 1995] while herb teas are generally low in tannins [Blake et al. 1993]. **Chlorogenic acid** is found in many fruits and vegetables, it is responsible for browning in apples and potatoes [Carper 1988]. Ascorbic acid may be added (eg. to apple juice) as an anti-oxidant to prevent browning.

Globe artichoke – *Cynara scolymus* (Asteraceae) contains the cinnamic acid depsides **cynarin** and **chlorogenic acid**.

Tannins as digestive inhibitors

Consumption of tannins may lead to reduced absorbtion of proteins and other nutrients. For this reason it is not considered wise to drink strong tea with meals. This can also lead to problems in compounding herbal medicines since there is a tendency to cause precipitates in some cases.

MISCELLANEOUS PHENOLIC COMPOUNDS

The basic phenolic structure occurs in many other classes of compounds found in medicinal herbs, including the following:

Glycosides eg. most flavonoids, anthraquinones
Essential oils eg. thymol
Alkaloids eg. oxyacanthine
Sterols

References

Adzek, T. & Camarasa, J. 1988. Pharmacokinetics of polyphenolic compounds. In *Herbs, Spices & Medicinal Plants Vol.3*. Oryx Press, Phoenix USA.

Ammon, H. & Wahl, M. 1990. Pharmacology of Curcuma longa. *Planta Medica* 57: 1-7.

Beckham, N. 1995. Phyto-oestrogens and compounds that affect oestrogen metabolism – Pt.1 *Aust. J. Med. Herbalism* 7:11-16.

Bombardelli, E. & Morazzoni, P. 1995. Vitis vinifera L. *Fitoterapia* LXVI: 291-317.

Blake, O. 1993/94. The tannin content of herbal teas. *Br J Phytotherapy* 3: 124-127.

Bruneton, J. 1995. *Pharmacognosy Phytochemistry Medicinal Plants*. Lavoisier Pubs, Paris.

Cadenas, E. & Hochstein, P. 1992. Pro- and antioxidant functions of quinones and quinone reductases in mammalian cells. *Advances in Enzymology* 65: 97-146.

Carper, J. 1988. *The Food Pharmacy*. Bantam Books, New York

Constantino L. et al. 1994. Composition, superoxide radicals scavenging and antilipoperoxidant activity of some edible fruits. *Fitoterapia* LXV: 44-47.

Dorsch, W. 1994. Antiasthmatic acetophenones – an in vivo study on structure activity relationship. *Phytomedicine* 1: 47-54.

Greinwald, R. & Stobernack, H. 1990. *Ammi visnaga* (Khella). *Br. J. Phytotherapy* 1: 7-10.

Haslam, E. et al. 1989. Traditional herbal medicines – the role of polyphenols. *Planta Medica* 55: 1-8.

Hikino, H. 1985. Chinese medicinal plants used against hepatitis. In Chang, H.M. et al. (eds) *Advances in Chinese Medicinal Materials Research*. World Scientific Pubs, Singapore.

Huang, Kee Chang 1993. *The Pharmacology of Chinese herbs*. CRC Press, USA.

Johri, J.K. et al. 1992. Coumarins as potent biocides... *Fitoterapia* LXIII: 78-80.

Kashiwada, Yoshiki et al. 1992. Antitumor agents, 129. Tannins and related compounds as selective cytotoxic agents. *J.Natural Products* 55(1992): 1033.

Kong, Y.C. et al. 1989. Antifertility Principle of *Ruta graveolens*. *Planta Medica* 55: 176-178.

Lazarova, G. et al. 1993. Photodynamic damage prevention by some hydroxycoumarins. *Fitoterapia* LXIV: 134-136.

Liviero, L. et. al. 1994. Antimutagenic activity of procyanidins from Vitis vinifera. *Fitoterapia* LXV : 203-209.

Lutete, T. et al. 1994. Antimicrobial activity of tannins. *Fitoterapia* LXV: 276-278.

MacRae, W. et al. 1989. The antiviral action of lignans. *Planta Medica* 55: 531-535.

Murakami, Shigeru et al. 1991. Inhibition of gastric H^+, K^+ -ATPase and acid secretion by ellagic acid. *Planta Medica* 57: 305-308.

Meier, B. & Liebi, M. 1990. Medicinal plants containing salicin: effectiveness and safety. *Br. J. Phytotherapy* 1: 36-42.

Okuda et al. 1988. Ellagitannins as active constituents of medicinal plants. *Planta Medica* 55:117-122.

San Feliciano, A. et al. 1993. Antineoplastic and antiviral activities of some cyclolignans. *Planta Medica* 59: 246-249.

Scalbert, A. 1991. Antimicrobial properties of tannins. *Phytochemistry* 30: 3875-3883.

Sur, P. & Ganguly, D. 1994. Tea plant root extract (TRE) as an antineoplastic agent. *Planta medica* 60: 1106-109.

Towers, G. 1980. Photosensitizers from plants and their photodynamic action. In Reinhold et al. (eds) *Progress in Phytochemistry Vol. 6*. Pergamon Press, Oxford.

Waterman, P. & Mole, S. 1994. *Analysis of Phenolic Plant Metabolites*. Blackwell Scientific Pubs, Oxford, UK.

Wohlmuth, H. 1995. Proanthocyanidins: a review of recent research into condensed tannins. *Proceedings of the 1995 NHAA International Conference*. NHAA, Sydney.

Chapter 3
GLYCOSIDES

INTRODUCTION

Glycosides are a group of compounds characterised by the fact that chemically they consist of a sugar portion (or moiety) attached by a special bond to one or more non-sugar portions. Chemically they are hydroxyls of a sugar that are capable of forming ethers with other alcohols.

Glycosides are broken down upon hydrolysis with enzymes or acids to:

1. sugar moiety = glycone
2. non-sugar moiety = aglycone / active portion. This may be a phenol, alcohol or sulphur compounds.

The bond between the two moieties may involve a phenolic hydroxly group in which case an O-glycoside is formed. Similarly carbon (C-glycosides), nitrogen (N-glycosides) or sulphur (S-glycosides) may be involved.

The linkage of the two moeities involves the transfer of a uridylyl group from uridine triphosphate to a sugar 1-phosphate. The glycoside is formed by the transfer of the sugar from uridine diphosphate to a suitable acceptor (aglycone), thus forming the glycoside [Tyler 1988].

Isolation of glycosides from plants is often difficult due to the tendency for plants to contain enzymes that bring about their own hydrolysis [Samuelsson 1992].

Distribution

Glycoside distribution is widespread throughout the plant kingdom. They occur in the seeds of pulses, swollen underground roots or shoots (yams, sweet potatoes), flowers and leaves.

Some may be toxic especially cyanogenic and cardiac glycosides. Cooking usually renders them non-toxic. They are mostly soluble in water and organic solvents, though the aglycones are somewhat less soluble.

Classification of glycosides is based on the nature of the aglycone. Despite the widespread distribution of some glycoside classes we often find that the same botanical families consistently contain the same aglycone types eg.

Brassicaceae – glucosinolate
Rosaceae – cyanogenic
Scrophulariaceae – cardiac, phenylpropanoid
Asteraceae – phenylpropanoid, flavonoid
Malvaceae – anthocyanin

CYANOGENIC GLYCOSIDES

In cyanogenic glycosides the element nitrogen occurs in the form of hydrocyanic acid , also known as prussic acid – one of the most toxic of all plant compounds. A number of amino acids are known precursors of these glycosides. **Amygdalin** from the bitter almond is derived from the aromatic amino acid phenylalanine. Amygdalin is hydrolysed in the presence of the enzyme amygdalase and water, involving a two stage process to produce glucose along with an aglycone made up of benzaldehyde (scent of bitter almonds) and odourless hydrocyanic acid.

amygdalin

The occurrence of cyanogenic glycosides is widespread. **Amygdalin** and **prunasin** are very common amongst plants of the Rosaceae, particularly the *Prunus* genus. This includes not only the bitter almond, but also the kernals of apricots, peaches, and plums. Marzipan flavour is also derived from amygdalin. Cyanogenic glycosides are found in other food plants, including linseed and manioc – a traditional flour in South America, where traditionally the plant is boiled and water discarded to remove the toxin. The glycosides are characteristic of several other plant families including the Poaceae (grasses) and Fabaceae (legumes).

sambunigrin

Sambunigrin (D-mandelonitrile glucoside) from the leaves of the Elder tree – *Sambucus nigra* (Caprifoliaceae) is isomeric to **prunasin**.

Toxicology

Toxicity of hydrocyanic acid involved inactivation of the respiratory enzymes, leading to dizziness and high facial colour. In high doses the whole of the central nervous system ceases to function and death follows. However massive doses of raw plant material (>3.5mg/kg) are required for a toxic effect to occur. Our bodies are able to neutralise cyanides by converting them to thiocynates which are eliminated in the urine [Bruneton 1995].

Therapeutics

An amygdalin containing drug called **laetrile** has been used as an anti-cancer medicine, though its use is now restricted. In small quantities these glycosides do exhibit expectorant, sedative and digestive properties. *Prunus serotina* or wild cherry bark is an excellent cough remedy and tonic, as well as a flavouring agent used in cough syrups. It is of benefit as a tea for bronchitis. The main active principle is **prunasin**.

CARDIAC GLYCOSIDES

These are glycosides possessing lactone rings attached in the β-position at C-17. Sugar residues are linked glycosidically via the C-3-OH groups of the steroid aglycones. The aglycones have a tetracyclic steroidal nucleus with hydroxyl groups at positions 3 and 14.

Most herbs that contain these compounds (and the compounds themselves) are scheduled for use by medical practitioners only, however all health practitioners should have a basic understanding of their pharmacology since they are so widely prescribed in general practice.

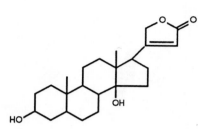

cardenolide

Biogenesis

Aglycones are derived from mevalonic acid, but the final molecules arise from a condensation of a C21 steroid with a C2 unit = cardenolides , C3 unit = bufadienolides

Sugar moeities are composed of 3 sugar units: glucose, rhamnose and specific sugars such as digitoxose which occur only in conjunction with cardiac glycosides. The sugar moeity confers on the glycoside solubility properties important in its absorbtion and distribution in the body. The presence of OH groups increases the onset of action and subsequent dissipation from the body. Glycosides with few OH groups tend to be lipophilic (fat soluble) and are absorbed and eliminated more slowly [Bruneton 1995]. The most widely used drug in this category is **digoxin**, which is actually a derivative of lantoside C, one of the glycosides in *Digitalis lanata* [Samuelsson 1992].

A select group of plant families are known to produce cardiac glycosides, the most notable being the Liliaceae, Scrophulariaceae and Apocynaceae. Some of the more important of the glycosides are listed along with their plant source:

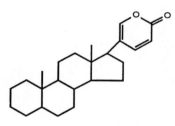

bufadienolide

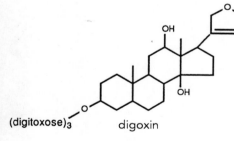

digoxin

digitoxin, gitoxin	Foxglove – *Digitalis purpurea* (Scrophulariaceae)
lanatosides A-E, digoxin	*Digitalis lanata*
convallotoxin, convalloside	Lily of the valley – *Convallaria majalis* (Liliaceae)
stropanthin	*Strophanthus gratus* (Apocynaceae)
hellebrin	Green hellebore – *Helleborus viridis* (Ranunculaceae)
proscillaridin; scillaren A	Squill – *Urginea maritima* (Liliaceae)
odoroside, oleandrin	Oleander – *Nerium oleander* (Apocynaceae)

Action of cardiac glycosides

Cardiac glycosides increase the force and speed of systolic contraction. In the failing heart they cause a more completely emptying of ventricles and shortening in length of systole. The heart has more time to rest between contractions. Increased cardiac output causes a lower heart rate and increases renal excretion. As to the pharmacology of digoxin itself, four main actions occur:

1. Positive inotropic effect – increases myocardial contractility due to direct inhibition of membrane bound Na+ K+ ATPase which leads to increased intracellular Ca++, ie. Ca++ replacing K+ leads to increase in muscle contraction.
2. Increase in atrial and ventricular myocardial excitability – may lead to arrythmias.
3. Decrease rate of atrioventricular conduction
4. Increase vagal tone and myocardial sensitivity to vagal impulses.

Toxicology

These compounds have a low therapeutic index (0.5), meaning the therapeutic dose is not much lower than the toxic dose. Digitalis intoxication affects the body in many ways. There are gastrointestinal symptoms (nausea, vomiting, diarrhoea), vision disturbances, neurological symptoms (headache, neuralgia, drowsiness) and cardiovascular symptoms including worsening cardiac failure and arrhythmias. Disturbance of electrolytes occurs and potassium may need to be administered. There are also many contraindications and drug interactions to be aware of. For a review of digoxin therapeutics and toxicology see Phillips & Johnston 1987.

PHENYLPROPANOID GLYCOSIDES

These glycosides (PhGs) have only been known of since 1964 when the first PhG, **verbascoside** – was isolated from *Verbascum sinuatum*. The structure of verbascoside was elucidated in 1968. Since then over 100 PhGs have been reported, however verbascoside is by far the most widespread, having been identified in more than 60 species from 14 plant families. These glycosides consist of 3 basic units:

1. a central glucose
2. a C6-C2 moiety, usually a dihydroxyphenyl-b-ethanol
3. a C6-C3 moiety, usually a hydroxycinnamic acid

phenylpropanoid glycoside

The aromatic units can be differentially derived and other saccharides are usually linked to 1 or 2 of the free hydroxyls of the central glucose. The precursors to the non-sugar moieties are tyrosine and cinnamic acid – products of the shikimic acid pathway. [Cometa et al 1993].

Phenylpropanoid glycosides are common in the following families:

Scrophulariaceae – *Verbascum* spp. *Rhemannia glutinosa; Digitalis purpurea*
Plantaginaceae – *Plantago asiatica*
Asteraceae – *Echinacea pallida; E.angustifolia*
Lamiaceae – *Stachys* spp. *Teucrium* spp.

Therapeutic actions of phenylpropanoids

- Antifungal
- Antibacterial eg. *Forsythia* spp. used in Oriental medicine; **echinocoside** in *Echinacea* spp.
- Antiviral esp. **echinocoside**, proven in animal studies.
- Immunosuppresive – inhibits HPFC [Haemolytic plaque-forming cells] esp. **verbascoside** in *Rehmannia glutinosa* and *Verbascum* spp.
- Antineoplastic – **verbascoside** inhibits PKC [Protein kinase C] involved in cellular proliferation and differation
- Platelet aggregation – PhGs from *Forsythia* spp. strong radical scavenging action

ANTHRAQUINONES

Also known as anthracene glycosides, since **anthracene** was the first compound isolated, by French chemists Dumas and Lambert, in 1832. Anthraquinones are yellow-brown pigments and many plants which contain them have a history of use as dyes for textiles eg. *Rubia tinctoria* – dyers madder. The aglycones consist of two or more phenols linked with a third carbon ring. Hydroxyl groups always occur at positions 1 and 8, hence they are 1,8 anthraquinones which form O-glycosides.

aloe-emodin

Experimental investigations with the most widely prescribed anthraquinones – **sennosides A & B** – show they pass through the stomach and small intestine unaltered, but that in the cecum and colon they are converted to dianthrones (their aglycones) by microorganisms. The **dianthrones**, which remain unabsorbed, are further transformed into anthrone and anthraquinone, producing hydragogue and laxative effects in the process [Adzet & Camarasa 1988]. The laxative effect is thought to occur as a result of increased peristalsis action and inhibition of water and electrolyte resorbtion by the intestinal mucosa. There is no evidence of direct irritation of the bowel mucosa [Bruneton 1995].

sennoside A

Therapeutic use of anthraquinones

The composition of glycosides and their derivatives in antraquinone containing plants determines their effectiveness as laxatives. The gentlest acting laxatives in this group belong to the buckthorns (*Rhamnus catharticus* and *R. frangula*) and rhubarb (*Rheum palmatum*). In both cases the herbs are aged for at least a year during which the more irritant anthraquinone derivatives are converted to milder acting compounds. The presence of tannins also tends to moderate the laxative effect.

Aloes (Aloe barbadensis) and *Senna* spp. are the other commonly used laxative agents in this class. Senna syrup is

hypericin

commonly prescribed for children and may be used during pregnancy and lactation for limited periods. Otherwise anthraquinones are contra-indicated during pregnancy. The duration of action is around 8 hours, and is usually taken before bed.

Due to the stimulant effect of these laxatives, they are contra-indicated in irritable/spastic colon conditions. A slight overdose can produce griping and discomfort, an effect which is generally counter-balanced by the presence of carminatives, eg. peppermint, coriander oil .

It is unwise to rely on these remedies alone when treating chronic constipation, since dependence can result. The anti-septic effects of anthraquinones deters the growth of enteric pathogens. Some anthraquinones and napthoquinones significantly inhibit Epstein-Barr virus early antigen activation at low doses [Konoshima et al 1989].

Hypericin, the dark red pigment from *Hypericum perforatum* is a dehydrodianthrone, structurally an anthraquinone. However it does not break down to anthraquinone in the bowel and is without laxative action. **Hypericin** has been thoroughly investigated and used (generally in *Hypericum* extracts standardised to hypericin content) for antidepressant and antiviral activities [Bombardelli & Morazzoni 1995].

FLAVONOIDS

These compounds occur as yellow and white plant pigments (Latin *flavus* = yellow). **Rutin** was discovered in rue (*Ruta graveolens*) in 1842 – it later became known as Vitamin P (permeability factor).

Chemistry
Flavonoids occur both in the free state and as glycosides. Their chemical structure is based on a C_{15} skeleton: C_6-C_3-C_6.

basic flavonoid structure
(with ring numbering system)

Flavonoids are products of both the shikimic acid and acetate pathways. Almost all plants studied have been shown to contain a series of closely inter-related flavonoids with different degrees of oxidation and/or hydroxylation patterns. While there are numerous structural classes within the flavonoids, the most commonly occurring are flavones (eg. **apigenin**) and flavonols (eg. **quercetin**).

apigenin

Another group of interest is the **anthocyanins**, responsible for the red, violet and blue colour of flowers and other plant parts. They are present in the plant as glycosides of hydroxylated 2-phenylbenzoprylium salts. **Cyanin** (cyanidin-3,5-diglucoside), is found in the cornflower – *Centaurea cyanus* [Asteraceae].

Flavonoids can be considered as important constituents of the human diet – average consumption is estimated at approximately 1g flavonoids/ person /day.

Role in plant physiology

Flavonoids are universal within the plant kingdom, they are the most common plant pigments next to chlorophyll and carotenoids. They are recognized as the pigments responsible for autumnal leaf colours as well as for the many shades of yellow, orange & red in flowers. Their function includes protection of plant tissues from damaging UV radiation, acting as antioxidants, enzyme inhibitors, pigments and light screens. The compounds are involved in photosentization & energy transfer, action of plant growth hormones & growth regulators, as well as defence against infection [Middleton 1988]. The plant response to injury results in increased synthesis of flavonoid aglycones (known as phytoalexins) at the site of injury.

quercetin

Therapeutics

Experiments have proven flavonoids affect the heart, circulatory system and strengthen the capillaries. They are often referred to as "biological stress modifiers" since they serve as protection against environmental stress [Middleton 1988]. They are also known to have synergistic effects with ascorbic acid. Their protective actions are mainly due to membrane stabilizing and anti-oxidant effects.

Enzyme inhibitors

The presence of aromatic hydroxyl groups & the fact that flavonoids can be broadly divided into 2 groups depending on differences in substitutions in the B-ring means the flavonoids are capable of alternately inhibiting or stimulating certain enzyme systems.

1. aldose reductase – causes diabetic cataracts eg. quercetrin, used as standard. Also luteolin, kampferol.
2. xanthine oxidase – causes hyperuricaemia which leads to gout, renal stones. It is postulated that free hydroxyl groups at C-5 & C-7 are important for inhibition of xanthine oxidase [Pathak et al. 1991]
3. tyrosine protein kinase – causes increased pre-malignant cells
4. lipoxygenase & cycloxygenase – involved with production of inflammatory prostaglandins, leukotrienes & thromboxanes ie. dual inhibitors . Arachidonic acid derivatives.
5. cyclic nucleotide phosphodiesterase (PDE) – key enzyme in promotion of platelet aggregation

Actions of flavonoids

1. Anti-inflammatory. Flavonoids such as **baicalein** from *Scutellaria baicalensis* [Lamiaceae] inhibit pro-inflammatory metabolites including certain prostaglandins and leukotrienes. **Quercetin** and others are effective inhibitors of the release of histamine induced by various agents. They inhibit a number of stages of inflammation including granulation tissue formation in chronic arthritis. Flavonoids offer the advantage of high margin of safety & lack of side effects such as ulcerogenicity, over the classical anti-inflammatory drugs.

2. Affect on capilliary permeability. Diseases associated with increased permeability of blood capillaries include diabetes, chronic venous insufficiency, haemorrhoids, scurvy, varicose ulcers, bruising. Buckwheat – *Fagopyrum esculentum* (Polygonaceae), contains 8% **rutin** if grown under suitable ecological conditions. Tissue examinations of animals with induced oedema plaques in corneal and conjunctival tissue showed that after treatment with 180mg rutin from buckwheat herb (tablets) the oedema was flushed out, the fragility of the vascular system was reduced. Photographic documentation shows quite distinctly the return to normal of the tissues after a period of 30 days [Schilcher& Muller 1981]

3. Anti-oxidant. Lipid peroxidation, the oxidative degradation of polyunsaturated fatty acids, is implicated in several pathological conditions – aging, hepatotoxicity, hemolysis, cancer, atherosclerosis, tumor promotion and inflammation. Selected flavonoids may exert protective effects against cell damage produced by lipid peroxidation stimulated by a variety of toxins, due to the antioxidative properties of the compounds. **Silymarin**, the flavono-lignan complex from *Silybum marianum*, protects liver mitochondria & microsomes from lipid peroxidation. This protection also occurs with the flavonoids **quercetin** and **taxifolin**.

Isoflavones

Isoflavones are flavonoid isomers whose distribution is restricted to the Fabaceae (legume) family. They have been classified with the flavonoids although they rarely occur in the glycosidic form. They have structural similarities to oestrogens so they are also classed amongst the phyto-oestrogens, compounds that bind to oestrogen receptors but whose oestrogen activity is relatively low. Isoflavones such as **genisten** found in soymilk and other soy products, help in prevention of breast and other tumors [Bone 1995]. Isoflavones are also found in liquorice (*Glycyrrhiza glabra*), alfalfa (*Medicago sativa*), and red clover (*Trifolium pratense*).

GLUCOSINOLATES
(Mustard oil glycosides)

These are pungent tasting compounds found mainly in the Brassicaceae family, though similar compounds are also present in the *Allium* genus (Liliaceae). While the *Allium* compounds occur less frequently as glycosides, they are usually classified along with the glucosinolates. These glycosides are formed by decarboxylation of amino acids such as tyrosine, phenylalanine and tryptophan. **Sinigrin** (potassium isothiocyanate), the glycoside from seeds of the black mustard seed (*Brassica nigra*), is hydrolised by the enzyme myrosin to the aglycone allyl isothiocyanate. Depending on conditions other thiocyanates and highly toxic nitriles may be formed, the latter when plants are subjected to very hot water (>45°C) [Mills 1994]. More than 70 individual glucosinolate compounds are known, varying only in the character of their side chain.

$$CH_2 = CH - CH_2 - C \big\langle {}^{S-glu}_{N-O-SO_3-K}$$

sinigrin

At least 300 species of Brassicas have been studied for their glucosinolate content. The compounds are mainly concentrated in the seeds, although they can be found anywhere in the plants. They can always be identified by their spicy, pungent taste – responsible for the flavours of mustard seeds, horseradish root, cress and rocket leaves etc. They also occur in the garden nasturtium – *Tropaeolum majus* (Tropaeolaceae), in the form of **glucotropaeolinoside**, which is hydrolized to the antibiotic compound benzyl isothiocynate.

Actions of glucosinolates.

Mustard oils act as rubefacients or irritants when applied topically, causing local vasodilation. Mustard poultices have been used historically to break up congestion in the lungs and bronchioles, though care must be taken not to induce skin lesions. Taken internally the compounds are effective decongestants for sinus conditions (eg. horse-radish & garlic tabs), while also acting to stimulate digestion. Large doses may induce emesis. As with all sulphur compounds they exhibit some antibiotic effects.

Glucosinolates also depress thyroid function, therefore all Brassicas are potentially goitregenic. Problems mainly arise when cattle or other livestock graze areas infested by *Brassica* weeds such as mustard weed or shepherds purse, and the milk or meat produced is consumed by local people who may also suffer from deficient iodine intake. Such occurrences have been recorded in some inland regions of Australia though there is no proof of a causal relationship [Bruneton 1995]. This goitregenic effect may also be harnessed therapeutically in cases of hyperthyroidism. The compounds are also responsible for some of the well documented antitumor properties of Brassicas [Brinker 1991]. Benzyl isothiocyanate, obtained by hydrolysis of glucosinolates in *Tropaeolum majus*, is cytotoxic and active against several human tumor cell lines [Pintao et al.1995].

IRIDOID GLYCOSIDES

Iridoids are synthesized through the mevalonic acid pathway and are technically known as cyclopentan-[c]-pyran monoterpenoids. They occur mainly as glycosides though non-glycosidic iridoids also occur – these are covered in the monoterpene section of the book. The name iridoid is derived from the common Australian meat ant *Iridomyrex detectus*, from which it was first detected in 1956 [Sticher 1977]. They are derived from plants belonging to many families, most notably the Rubiaceae, Lamiaceae, Scrophulariaceae and Gentianaceae.

Herbs containing iridoid compounds
The first of these compounds to be identified was **asperuloside** from the woodruff – *Asperula odorata* (Rubiaceae). Other iridoids with therapeutic properties include:

harpagoside

aucubin from plantain – *Plantago* spp. (Plantaginaceae)

harpagoside, procumbin from devil's claw – *Harpagophytum procumbens* (Pedaliaceae)

harpagoside occurs also in figwort *Scrophularia nodosa* (Scrophulariaceae)

loganin from bogbean *Menyanthes* spp. (Menyanthaceae)

Secoiridoids

These glycosides are formed by opening of the 5 carbon ring of the iridoid **loganin**. They include:

amarogentin, **gentiopicroside** from gentian – *Gentiana* spp. (Gentianaceae)

picroliv from *Picrorrhiza kurroa* (Scrophulariaceae)

oleuropein from olive leaves – *Olea europea* (Oleaceae)

gentiopicroside

Therapeutics of iridoid glycosides

Iridoids are the most bitter of all plant compounds, often responsible for the so called "bitter principle". On a scale for bitter value devised by Wagner & Vaserian [described in Sticher 1977], **amarogentin** and related secoiridoids were the most bitter compounds tested. Its taste is perceptible at a dilution of 1 part/ 50,000. Bitters are known to stimulate release of gastrin in the GIT, leading to increase in digestive secretions including bile flow.

Bitters improve appetite and assist pancreatic function. They are regarded as cooling remedies, useful for fevers and inflammations. Actions include:

Antiinflammatory – **aucubin**, **loganin** [Recio et al 1994].

Hepatoprotective – **picroliv** [Visen et al 1993], gentiopicroside, aucubin.

Hypotensive – **oleuropein** (also coronary dilating, antiarrhythmic)

Laxative – all iridoid compounds tested laxative in mice experiments. The strongest of those tested were only $1/7$ as potent as sennosides, however the iridoids were active in a very short time by comparison to the sennosides [Sticher 1977].

33

References

Adzet, T. & Camarasa, J. 1988. Pharmacokinetics of Polyphenolic Compounds. In Cracker & Simon [eds] *Herbs, Spices & Medicinal Plants Vol.3*. Oryx Press, Arizona.

Bisset, N. Houghton & Hylands. 1991. Flavonoids as anti-inflammatory plants. In R. Wijesekera. *The Medicinal Plant Industry*. CRC Press, Florida.

Bone, K. 1995. Oestrogen modulation. *The Modern Phytotherapist* 1 : 8-10.

Bombardelli, E. & Morazzoni, P. 1995. *Hypericum perforatum*. *Fitoterapia* LXVI: 43-68.

Brinker, F.1991. Inhibition of endochrine function by botanical agents II. *J.Naturopath. Med. 2*:18-22.

Bruneton, J. 1995. *Pharmacognosy Phytochemistry Medicinal Plants*. Lavoisier Pubs, Paris

Cometa, F. et al. 1993. Phenylpropanoid glycosides. Distribution & Pharmacological activity. *Fitoterapia* Vol. LXIV: 195-217.

Konoshima, T.et al. 1989. Studies on inhibitors of skin tumor promotion. *J. Nat. Products* 52: 987:995.

Middleton Jr., E. 1988. Plant flavonoid effects on mammalian cell systems. In Cracker & Simon [eds] *Herbs, Spices & Medicinal Plants Vol.3*. Oryx Press, Arizona.

Pathak, D. Pathak & Gingla. 1991 Flavonoids as medicinal agents – Recent Advances. *Fitoterapia* Vol. LXII: 371-389.

Phillips, P. & Johnston, C. 1987. Optimal use of cardiac glycosides. *Current Therapeutics*. June

Pintao, A. Pais, M. Coley, H. & Judson, I. 1995. *In vitro* and *in vivo* antitumor activity of benzyl isothyanate: a natural product from *Tropaeolum majus*. *Planta Medica* 61: 233-236.

Recio, M. Giner, R. Manez, S & Rios, J. 1994. Structural considerations on the iridoids as anti-inflammatory agents. *Planta Medica* 60: 232-234.

Reuter, H. 1995. *Allium sativum and Allium ursinum*: Part 2, Pharmacology & medicinal application. *Phytomedicine* 2: 73-91.

Samuelsson, G. 1992. *Drugs of Natural Origin*. Swedish Pharmaceutical Press, Stockholm.

Sticher, O. 1977. Plant mono-, di- and sesquiterpenoids with Pharmacological or therapeutic activity. In Wagner, H. & Wolff, P. [eds] *New Natural Products with Pharmacological, Biological or Therapeutic Activity*. Springer-Verlag, Berlin.

Vaughan, J. Macleod, A. & Jones, B. 1976. *The Biology & Chemistry of the Cruciferae*. Academic Press, London.

Chapter 4
POLYSACCHARIDES

Polysaccharides or glycans are high molecular weight polymers consisting of chains of sugars (mono- or oligo-saccharides) with chemical linkages. The simplest polysaccharides are cellulose and starch which are polymers of glucose only. Polysaccharides are of widespread distribution throughout the plant kingdom including the algaes, and also occur in the fungal kingdom. Their functions include food storage, protection of membranes, and maintaining rigidity of cell walls in plants, except for seaweeds where they help maintain the flexibility required for life in the ocean [Bruneton 1995].

GUMS and MUCILAGES

Chemistry

Compounds made up of branched chain of chemically linked sugars (monosaccharides) or their salts, and uronic acid derivatives (uronic acids are oxidation products of sugars). The gums are usually heterogenous ie composed of various monosaccharide residues and uronic acids. They may have acidic, basic or neutral characteristics. A typical example is **alginic acid** found in the seaweed bladderwrack or kelp – *Fucus vesiculosis* (Fucaceae) which yields mannuronic and guluronic acid residues.

β-D-mannuronic acid resides in alginic acid

Origins

Gums

Many plants (esp. growing in semi-arid conditions) produce gummy exudates when the bark is damaged – these serve to heal the wound. The exudate often dries to a hard amorphous mass, and is produced in sufficient abundance by some species of trees and shrubs to warrant collection and commercial utilization. Gum exudates are readily obtainable in relatively pure, undegraded form though in some cases purification is still required. These polysaccharides often occur in association with a protein.

Examples: **gum arabic** from *Acacia* spp. (Mimosaceae) ; **gum tragacanthe** – *Astragalus* spp. (Fabaceae); **Prunus** gums – *Prunus* spp. (Rosaceae).

Seed gums

These are obtained from certain types of seeds, particularly in their endosperms. They are common in the legume (Fabaceae) family eg: **Locust bean gum; tamarind** – *Tamarindus indicus*; guar gum. Their properties are similar to gum exudates.

Seaweed gums

Found in the leaves of several groups of algae eg. **agar agar** – red algae; **kelp** – brown algae; **Irish moss** (carrageenan) – red algae.

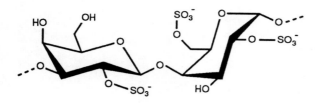

λ-carrageenan

Mucilages

These are slimy, semi-solid substances found in specialised cells in plants. They may be found in –

roots: comfrey	– *Symphytum officinalis* (Boraginaceae)
leaves: coltsfoot	– *Tussilago farfara* (Asteraceae)
bark: slippery elm	– *Ulmus fulva* (Ulmaceae)
seeds: psyllium	– *Plantago ovata, P. Psyllium* (Plantaginaceae).

Their properties are similar to those of seaweed gums and the term is often applied to aqueous suspensions or solutions of gums and gelatinised starches.

Properties of gums and mucilages

Hydrophilic (attract water) and, in energetic terms – cooling, sweet. Their physical properties are of more significance than their chemical properties. They are quite indigestable, however even if polysaccharides were broken down in the digestive tract, the breakdown products, sugars and uronic acids, have little pharmacological effect (Mills 1994). Therefore these compounds are sometimes regarded as inert. They are often considered undesireable and omitted from pharmaceutical preparations, except where 'gummy' properties are required eg. tablets, lozenges.

Solubility

Polysaccharides are insoluble in alcohol – they precipitate from alcohol based solvents.

Tinctures, which are made on alcoholic solvents of 45% strength or higher, are therefore of little use where demulcent or emollient effects are required.

The degree of water solublility depends on the polysaccharide structure. Linear polymers (mucilages) are less water soluble, they tend to precipitate at high temperatures and form viscous solutions. They may be referred to as 'slimy'. Branched polymers (gums) are more water soluble, forming gels. These can be referred to as 'tacky'[Mills 1994].

Local effects

The primary action is local ie. by direct contact with surface of mucous membranes or skin. Here they produce a coating of slime that acts to soothe and protect exposed or irritated surfaces of GIT – a demulcent action. When this effect occurs on the skin it is referred to as an emollient action.

Gums and mucilages are an invaluable aid in the management of irritable digestive disorders, especially where ulceration is a feature. Their relative indigestibility and hydrophilic properties create an influence on bowel behaviour:

1. laxative – bulk effect leads to peristalsis. Therefore high fibre diets are effective.
2. anti-diarrhoeal – small quantities absorb excess water in colon. Tannins if present have a binding effect eg. slippery elm powder.

They are ideal for treatment of irritable bowel disorders, especially psyllium husk. **Alginates** from seaweeds are used in pharmaceutical gastric antacids for relief of gastric reflux, oesophagitis and hiatal hernias [Bruneton 1995]. Slippery elm powder is equally effective for these disorders.

Mucilage retains a large amount of water and hence maintains an elevated temperature, which penetrates the tissues progressively – especially as hot compress eg. linseed. It also checks fermentation and bacterial growth, adsorbs toxins and wastes helping their elimination from the body. Cholesterol is also lowered through this mechanism – a general property of water soluble fibres such as those in oat bran. The hydrophilic effect produces a sensation of 'fullness' in the stomach without providing calories, hence the use of guar gum and others for appetite suppression. There is also a blood sugar lowering effect which has been observed in both diabetics and normal subjects.

Reflex effects

The soothing, demulcent effects of gums and mucilages also benefit irritable states of the urinary and respiratory tract. Take for example the widespread use of mucilagenous herbs such as *Althaea officinalis* or *Symphytum officinale* for bronchitis, and soothing preparations such as barley water for cystitis. While there is little difficulty in comprehending the strictly localised demulcent effect on the lining of the digestive tract, the mode of operation on organs with which the mucilage doesn't come in contact – respiratory tract, genito-urinary tract and uterus – is less clear. In recent years a new theory has emerged, suggesting the effects occur via reflex associations with the digestive tract, through an embryonic link in the nervous system [Mills 1994].

IMMUNOSTIMULATING POLYSACCHARIDES

These are a group of water soluble, acidic, branch-chained polysaccharides with very high molecular weights. They may be bound to proteins. It has been observed that immuno-stimulating activity is highest in compounds which contain β1,3 linkages at position 6 [Willard 1992].

Mushroom polysaccharides

Polysaccharides are widely distributed in both the plant and the fungal kingdoms. For immunostimulating function to occur mushroom polysaccharides must contain uronic acid and be of high molecular weight [Srivastava & Kulshreshtha 1989]. Studies have shown most of the polysaccharide rich medicinal mushrooms used in China suppress sacroma 180 tumors in mice [Ying et al. 1987]. One of these, **lentinan**, found in the shiitake mushroom –

Lentinus edodes (Tricholomataceae), is a potent T lymphocyte restorer in cancer patients, and has an inhibiting effect on the HIV virus.

lentinan

Lentinan possesses anticomplementary activity ie enhances immunity by activation of the alternative complement pathway [Srivastava & Kulshreshtha 1989]. In common with some other antitumor polysaccharides, **lentinan** is taken intravenously to induce its antitumor effects [Werbach & Murray 1994]. However in a study on the effects of mushroom polysaccharides administered orally on transplanted tumors in mice, lentinan from shiitake caused up to 50% of tumors to regress, while a β-glucan polysaccharide from maitake mushroom – *Grifola frondosa* (Polyporaceae), also taken orally, caused over 80% regression of tumors [Nanba 1993]. The effectiveness of maitake polysaccharide (taken orally) in reversing breast, lung and liver cancers was recently confirmed in a clinical trial in Japan [Nanba 1996].

Similar polysaccharides are found in the reishi mushroom – *Gandoderma lucida* (Polyporaceae), one of the most potent herbal medicines for treating conditions of chronic fatigue and weakened immunity. *Gandoderma* contains numerous immunostimulating polysaccharides, mostly of the β-**D-glucan** type [Willard 1990] along with **gandodernans**, which have hypoglycaemic action [Hikino et al. 1989]. *Gandoderma* is also rich in triterpenoids and is classed as one of the great oriental 'adaptagen' medicines.

Polysaccharides from ginseng – *Panax ginseng* (Arailaceae) show anti-ulcerogenic activity against induced ulcers [Sun et al.1991], while the pectin-like **bupleuran 2II** from *Bupleurum falcatum* (Apiaceae) exhibited significant anti-ulcer activity [Yamada et al 1991].

MISCELLANEOUS POLYSACCHARIDES

Pectins
Pectins are complex galactouronic acid-based carbohydrates found in the plant cell wall of many fruits eg. apples, citrus. During fruit ripening an insoluble precursor is converted to soluble pectin and it becomes gelatinous. It is used as setting agents for jams. Pectin has similar properties to gums – it adsorbs toxins, cholesterol and acts as a bulk laxative agent.

Inulin or Fructans
Fructans are polymers of fructose stored in some plants as reserve material instead of starch. They have much lower molecular weight than starch, and are water soluble. The branched fructans are found mainly in the grass family (Poaceae) while linear fructans (specifically inulin) are particularly common in the Asteraceae. **Inulin** contains 35 fructose residues along with a terminal glucose.

Inulin helps stabilise blood sugar in hypoglycaemia, and also has diuretic and immunostimulating properties. It is found in significant quantities in the following herbs:

Elecampane – *Inula helenium*
Burdock – *Arctium lappa*
Globe artichoke – *Cynara scolymus*
Chicory – *Cichorium intybus*
Dandelion – *Taraxacum officinalis*
Echinacea spp.

inulin (piece of the polysaccharide chain)

References

Fluck, H. 1976. *Medicinal Plants.* W. Foulsham & Co, U.K.

Hikino, H. et al. 1989. Mechanisms of hypoglycemic activity of ganoderan B: a glycan of *Gandoderma lucidum* fruit bodies. *Planta Medica* 55: 423428.

Miller, L. [ed]. 1973. *Phytochemistry Vol.1* Van Nostrand Reinhold Cc, New York

Mills, S. 1994 . *The Essential Book of Herbal Medicine.* Viking Arkana, London.

Nanba, H.1993. Antitumor activity of orally administered "D-fraction' from Maitake mushroom (*Grifola frondosa*). *J. Naturopathic Med.* 4: 10-15.

Nanba, H. 1996. Maitake D-fraction. Healing and preventing potentials for cancer. *Townsend Letter for Doctors and patients.* Feb/March.

Srivastava, R. & Kilshreshtha, D. 1989. Bioactive polysaccharides from plants. *Phytochemistry* 28: 2877-2883.

Sun, X. Matsumoto, T. & Yamada, H. 1992. Anti-ulcer activity and mode of action of the polysaccharide fraction from the leaves of *Panax ginseng. Planta Medica* 58: 432-435.

Werbach, Melvyn & Murray, Michael. 1994. *Botanical influences on illness. A sourcebook of clinical research.* Third Line Press, Tarzana, California.

Willard, T. 1990. *Reishi mushroom.* Sylvan Press, Washington.

Willard, T. 1992. *Textbook of Advanced Herbology.* Wild Rose College of Natural Healing, Alberta.

Yamada, H. Sun, X. Matsumoto, T. Ra, K. Hirano, M. & Kiyohara, H. 1991. Purification of anti-ulcer polysaccharides from the roots of *Bupleurum falcatum. Planta Medica* 57: 555-559.

Ying, J. Mao, X. Ma, Q. Zong, Y. & Wen, H. 1987. *Icons of medicinal fungi from China.* Science Press,Beijing.

Chapter 5
TERPENOIDS & SAPONINS

TERPENOIDS

Terpenoids or terpenes comprise one of the most important group of active compounds in plants with over 20,000 known structures. All terpenoid structures may be divided into isoprene units, which arise from acetate via the mevalonic acid pathway.

Isoprenes are 5-carbon units (hydrocarbons) containing 2 unsaturated bonds.

$$CH_3$$
$$|$$
$$CH_2 = C — CH = CH_2$$

Isoprene $C_5 H_8$
Terpene $(C_5 H_8)n$

During the formation of the terpenoids, the isoprene units are linked in a head to tail fashion. The number of units incorporated into a particular terpene serves as a basis for the classification of these compounds.

Monoterpenes	$C_{10} H_{16}$	Essential oils eg. menthol. Iridoids
Sesquiterpenes	$C_{15} H_{24}$	Bitter principles esp. sesq. lactones
Diterpenes	$C_{20} H_{32}$	Resin acids, bitter principles
Triterpenes	$C_{30} H_{48}$	saponins, steroids
Tetraterpenes	$C_{40} H_{64}$	carotenoids
Polyterpenes	$(C_5 H_8)n$	rubber

MONOTERPENES

These are the major class of chemical compounds found in essential oils. Among the most widely occurring monoterpene oils are cineole from *Eucalyptus* spp. and pinene from *Pinus* spp.

Biosynthesis involves condensation of two C_5 precursors giving rise to geranyl pyrophosphate, a C10 intermediate from which monoterpenes are derived [Harborne & Baxter 1993]. Condensation of geranyl pyrophosphate with C_5 precursors gives rise to diterpenes, and the other classes of terpenoids are similarly formed.

Iridoids

The bitter iridoid monoterpenes usually occur as glycosides. Nonglycosidic iridoids include the sedative **valepotriates** found in valerian – *Valeriana* spp. (Valerianaceae). They are of interest chemically because they represent a structural link between terpenes and alkaloids. They are closely related to **nepetalactone**, the volatile component of essential oil of catnip –*Nepeta cataria* (Lamiaceae) which is responsible for attracting cats and other animals to the plant.

Paeoniflorin, a monoterpene glucoside, is a major constituent found in the Chinese herb *Paeonia lactiflora* (Ranunculaceae). The herb is antiinflammatory, sedative, antipyretic and antispasmodic [Huang 1993].

nepetalactone

SESQUITERPENES

These 15C compounds occur mainly as ingredients of essential oils or as γ-lactones. They are thought to have evolved as phytoalexins or antifeedants, compounds synthesized by plants as a response to fungal attack and herbivore grazing [Cronquist 1988; Bruneton 1995].

Gossypol from the cotton plant *Gossypium herbaceum* (Malvaceae) is a sesquiterpene dimer, consisting of two bonded napthalene-like structures.

Gossypol has been used in China as an antifertility agent in men. It is thought to act primarily on the testicular mitochondria, by inhibiting Ca+ uptake at

gossypol

the presynaptic endings [Huang 1993]. Despite the high efficacy (> 99%) of gossypol its long term use as an infertility agent is limited by undesireable side effects which include irreversible infertility [Hamburger & Hostettmann 1991].

Oil derived from the German chamomille – *Chamomilla recutita* (Asteraceae) contains the blue coloured sesquiterpene **chamazulene** as well as the anti-inflammatory compounds, **bisabolol** and **bisabolol oxides**.

chamazulene

SESQUITERPENES LACTONES

These bitter compounds contain 3,000 different structures. Although they can be found in a few fungi and other Angiosperm families such as Apiaceae and Lauraceae, they are most characteristic of herbs of the Asteraceae family [Bruneton 1995]. They tend to concentrate in the leaves and flowers, often as mixtures of several related compounds [Harborne & Baxter 1993]. The suffix "olide" indicates the presence of a lactone group.

(-)-α-bisabolol

Several Asteraceous medicinal plants and their main sesquiterpene ingredient are listed below:

Achillea millefolia – achillin
Inula helenium – alantolactone
Artemisia annua – absinthin
Cnicus benedictus – cnicin
Arnica montana – helenalin
Tanacetum parthenium – parthenolide

alantolactone

Therapeutic actions

Antiinflammatory	– helenalin
Antitumor	– parthenolide, helenalin [Rodriguez et al. 1976].
Stomachic	– cnicin, absinthin
Antibiotic	– alanolactone [Boatto et al 1994].
Migrane preventative	– parthenolide

Allergic contact dermatitis

Numerous sesquiterpene lactone containing plants of the Asteraceae are known to cause contact dermatitis in humans. In a recent case a patient developed acute dermatitis of the right hand with severe blistering following a single application of Arnica tincture [Hormann & Korting 1995]. The presence of a α-methylene group in the chemical structure is thought to be a pre-requisite for production of dermatitis [Rodriguez et al. 1976].

DITERPENES

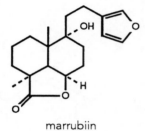

marrubiin

These are the most bitter tasting of all terpenoid compounds, responsible for the acclaimed stomachic and tonic properties of herbs such as **marrubiin** in horehound – *Marrubium vulgare* (Lamiaceae) and **calumbin** from calumba – *Jatrorrhiza palmata* (Berberidaceae).

Diterpenes tend to be most abundant in the Lamiaceae family, eg. *Salvia officinalis* or common sage which has antiviral diterpenes [Tada et al. 1994]. Some diterpenes have been found to have profound effects for a range of medical disorders. They include:

Taxol

Pacific Yew – *Taxus baccata* (Taxaceae). Prescription drug for ovarian cancer and other tumors [Lenaz & De Furia 1993]. The antimitotic effect is thought to be related to the interaction of taxol with the tubulin-microtubule system [Hamburger & Hostettmann 1991].

taxol

Forskolin

Coleus forskohlii (Lamiaceae). Vasodilatory; antihypertensive; bronchodilatory; positive inotropic action on heart; decreases intraocular pressure; inhibits platelet aggregation [Hamburger & Hostettmann 1991; Bruneton 1995].

Andrographolide

Andrographis paniculata (Acanthaceae). Antihepatoxic [Chander et al.1995].

TRITERPENES

Triterpenoid compounds are derived from a C_{30} precursor, squalene, which was first isolated from shark liver [Bruneton 1995]. They have similar configurations to steroids (found in plants and animals) whose C_{27} skeletons are also derived from squalene.

The triterpenoids are a large and diverse group made up of several sub-classes:

- Triterpenoid saponins
- Steroidal saponins
- Cardiac glycosides
- Phytosterols
- Curcurbitacins
- Quassinoids

triterpenoid notational system

PHYTOSTEROLS

Sterols such as **stigmasterol** and **sitosterol** are essential components of cell membranes, and they are also used as the starting material in the production of steroidal drugs. Phytosterols are characterized by a hydroxyl group attached at C_3 and an extra methyl or ethyl substituent in the side chain which is not present in animal sterols [Harborne & Baxter 1993].

stigmasterol

Phytosterols are minor but beneficial components of the human diet since they may inhibit growth of tumors and help in regulation of blood cholesterol. Therapeutically they are important constituents of the following herbs:

Withania somniferum (Solanaceae), known as ashwaganda in Ayurvedic medicine, contains steroidal lactones called **withanolides** which exhibit antitumor and hepatoprotective activities.

Urtica dioica – stinging nettle root. Recent experiments demonstrate a potent inhibition of enzymes involved in benign prostatic hyperplasia. Steroidal compounds including **stigmast-4-en-3-one** are thought to be responsible for this activity [Hirano et al 1994].

Commiphora mukul – myrrh (known as guggal in India). The resin contains steroids known as **guggulsterones** which lower blood cholesterol and triglycerides via stimulation of thyroid function [Bruneton 1995].

SAPONINS

Saponins are naturally occurring glycosides whose active portions are soluble in water, and which produces a lather. The combination of lipophilic aglycones (or sapogenins) with water-soluble sugars gives them the ability to lower surface tension, producing the characteristic detergent or soap-like effect on membranes and skin [Harborne & Baxter 1993]. Classification of saponins is based on the chemical structure of their aglycones, referred to as sapogenins.

Triterpenoid saponins

The most widely distributed aglycone is **oleanolic acid** – hence the "oleanolic acid ring system". Other aglycones have their own characteristic structures, eg. the **ursane** and **dammarane** ring systems. This group is pentacyclic ie. contains five carbon rings.

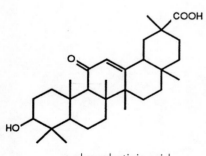

glycyrrhetinic acid

Steroidal saponins

These saponins contain steroid structures as their aglycones. They possess C_{27} skeletons in structures which are generally arranged into six rings. Some are used as precusors of sex hormones, cortisone, vitamin D. Examples are **diosgenin** and **hecogenin**.

diosgenin

Glycoalkaloids

A sub-group of the steroidal saponins are the glycoalkaloids, in which the aglycone is a steroidal alkaloid (contains an N atom). These are commonly found in the *Solanum* genus.

solasodine

THERAPEUTIC ACTION OF SAPONINS

Local effect

The detergent effect referred to above is important in understanding the significant local effects of saponins. Many good wound healing herbs are rich in saponins, including *Calendula officinalis* and *Arnica montana* (both Asteraceae). Irritant effect on mucous membranes of respiratory and urinary tracts is responsible for the expectorant and diuretic properties of saponin rich herbs such as *Polygala senega* (Polygonaceae) and *Primula* spp. (Primulaceae).

Their local effect is also observed in the digestive system where they increase and accelerate the bodies ability to absorb some active components eg. calcium and silica from foods. They generally assist digestion (though can become emetics in large quantities) and are found in many foods including tomatoes, spinach, asparagus, soya beans and oats.

Systemic effect

The aglycones once absorbed have favourable effects on blood vessel walls. Saponins from the horse chestnut – *Aesculus hippocastanum* (Hippocastanaceae) and butchers broom – *Ruscus aculeatus* (Liliaceae) are beneficial for varicose veins and haemorrhoids, while those from lime flowers – *Tilea* spp.(Tiliaceae) assist by lowering blood cholesterol.

Haemolytic effect

Saponin aglycones are capable of increasing permeability of membranes. They can cause haemolysis by destroying the membranes of red blood cells, thus releasing the haemoglobin.

Taken orally the detergent effect is nullified in the stomach before aglycones are absorbed into the bloodstream eg. *Phytolacca decandra* (Phytolaccaceae). Haemolysis can also occur by absorbtion through skin – cuts, eyes etc. A recent study revealed haemolytic rates in steroidal saponins were fast compared to those of triterpenoids. [Takechi & Tanaka 1995].

Other actions of saponins

Anti-inflammatory – **glycyrhizzin** from liquorice
Antifungal, antimicrobial
Hepatoprotective – **saikosaponins** from *Bupleurum* spp.
   (Apiaceae), **ginsenosides** from *Panax* spp. (Araliaceae).
Molluscicidal

The chemical composition of some saponins is very similar to that of human hormones.

Their aglycones have similar chemical structure to steroid hormones ie. stress & sex hormones from adrenals and gonads. They form the chemical basis for the herbal adaptagens.

Herbal adaptagens

These agents are referred to as 'harmony remedies' by Stephen Fulder in his excellent book originally titled "The Root of Being. Ginseng and the Pharmacology of Harmony" [Fulder 1980], still one of the best analyses of this class of medicinal agent. The concept of adaptagens is based around enhancement of vitality and general resistance rather than treatment of specific illnesses.

" By helping the body cope with stress, adaptogens can help accelerate learning speed, improve the memory, increase stamina in high performance athletes, alleviate small complaints and cut down infections by acting as a prophylactic" [Wahlstrom 1987].

Other compounds besides saponins may also be implicated in adaptagenic activity, particularly polysaccharides such as those found in some well known medicinal mushrooms.

Examples of adaptagenic herbs

Panax ginseng and *Panax* spp., *Eleuthrococcus, Glycyrhizza, Smilax* spp., *Dioscorea* spp., *Centella asiatica, Bupleurum* spp., and *Gandoderma* – the reishi mushroom.

Major saponin containing herbs

Expectorants: Golden rod, poke root, mullein, squill, violet leaves, liquorice, soapwort, senega, bittersweet.

Diuretics: Corn silk, *Primula*, scarlet pimpernell.

Anti-inflammatories: Wild yam, chickweed, sarsaparilla, liquorice, Yucca, *Bupleurum*

References

Boatto, G. et al. 1994. Composition & antibacterial activity of *Inula helenium* and *Rosmarinus officinalis* essential oils. *Fitoterapia* LXV: 279-280.

Bruneton, J. 1995. *Pharmacognosy Phytochemistry Medicinal Plants.* Lavoisier Pubs, Paris

Chander, Ramesh et al. 1995. Antihepatoxic activity of diterpenes of *Andrographis paniculata. Int. J. Pharmacognosy* 33: 135-138.

Della Loggia, R. Tubaro, A. Sosa, S. Becker, H. Saar, St. & Isaac, O. 1994. The role of triterpenoids in the topical antiinflammatiory activity of *Calendula officinalis* flowers. *Planta Medica* 60: 516-520.

Favel, A. et.al. 1994 *Planta Medica* 60: 50.

Fulder, S. 1980. *The Root of Being.* Hutchinson, London. (Also published as "*The Tao of Medicine*").

Hamburger, M. & Hostettmann, K. 1991. Bioactivity in plants: the link between phytochemistry and medicine. *Phtyochemistry* 30: 3864-3874.

Harborne, J. & Baxter, H. 1993. *Phytochemical Dictionary.* Taylor & Francis, London.

Hirano, T. Homma, M. & Oka, K. 1994. Effects of stinging nettle root extracts and their steroidal components on the Na+, K+ - ATP ase of the benign prostatic hyperplasia. *Planta Medica* 60: 30-33.

Hormann, H. & Korting, H. 1995. Allergic acute dermatitis due to *Arnica* tincture self-medication. *Phytomedicine* 4: 315-317.

Huang, K. 1993. *The Pharmacology of Chinese herbs.* CRC Press, USA.

Lenaz, L. & De Furia, M.D. 1993. Taxol: a novel natural product with significant anticancer activity. *Fitoterapia* LXIV: 27-36.

Ming-Hong Ye, et al. Anti-inflammatory and hepatoprotective activity of saikosaponin-f and the root extract of *Bupleurum kaoi*. *Fitoterapia* Vol.LXV: 409-412.

Nicholas, H.J. 1973. Terpenes. In L. Miller (ed) *Phytochemistry* 2. Van Nostrand, New York.

Rodriguez, E. Towers, G. & Mitchell, J. 1976. Biological activities of sesquiterpene lactones. *Phytochemistry* 15: 1573-1580.

Sticher, O. 1977. Plant mono-, di- and sesquiterpenoids with Pharmacological or therapeutic activity. In Wagner, H. & Wolff, P. [eds] *New Natural Products with Pharmacological, Biological or Therapeutic Activity.* Springer-Verlag, Berlin.

Tada, M. 1994. Antiviral diterpenes from *Salvia officinalis*. *Phytochemistry* 35: 539-541.

Takechi, M. & Tanaka, Y. 1995. Haemolytic time course differences between steroid and triterpenoid saponins. *Planta Medica* 61: 76-77.

Ushio & Abe. 1991. *Planta Medica* 57:511.

Wagner, H. Bladt, S. & Zgainski, E. 1984. *Plant Drug Analysis.* Springer-Verlag. Berlin.

Wahlstrom, M. 1987. *Adaptogens. Natures key to well-being.* Utgivare, Goteborg.

Willard, T. 1992. *Textbook of Advanced Herbology.* Wild Rose College of Natural Healing, Alberta.

Worth, H. & Curnow, D. 1980. *Metabolic pathways in medicine.* Edward Arnold, London.

Chapter 6
OILS & RESINS

ESSENTIAL OILS

Essential oils are odourous principles which are stored in special plant cells eg. glands, glandular hairs, oil ducts and resin ducts.They may occur in flowers, fruits, leaves, roots, wood, stems, bark and saps. These oils are responsible for the distinctive aromas associated with individual plant species. They are soluble in alcohol and fats, but only slightly soluble in water. Most essential oils are colourless, apart from **azulene** which is blue. On exposure to light and air they readily oxidise and resinify. They are also called volatile oils, since they evaporate when subjected to heat.

Extraction of oils
Steam distillation is the predominent extraction method used. Other methods include solvent extraction, cold pressing, infusion, effleurage and water distillation. Distillation is a method that involves the evaporation and subsequent condensation of liquids in order to produce, refine, and concentrate essential oils. High quality oils are distilled once only, whilst some commercial oils are "purified" by double or triple distillation methods.

The Olfactory System
This is the structure which is responsible for our sense of smell. The olfactory nerves connect directly to the limbic system in the brain, and thereby influence sensory functions such as hunger, sex and emotions. Smelling involves the inhalation of microscopic chemicals such as those contained in essential oils, which flow through our nostrils into the nasal cavity. Here they pass over moist bony structures called turbinates to reach the olfactory receptor cells, where the chemical dissolves and comes in contact with a fine layer of hairs, which then stimulates the

olfactory bulb of the brain [Wrigley & Fagg 1990]. Research into the chemistry of plant odours shows there is a definite link between the shape of molecules and their smell. Compounds that occur as optical isomers (enantiomers) produce odours that differ according to which isomer is present, so that compounds with the same chemical structures can have different odours [Pavia et al. 1982].

Chemotypes of oils

Essential oil composition can vary according to geographic and genetic factors, even though the same botanic species is involved – a phenonemon known as chemical polymorphism. When this occurs a terminology can be used where the Latin name is followed by the name of the chemical component most characteristic for that particular race of the plant ie. its chemotype –

> eg. *Thymus vulgaris* linalol
> *Thymus vulgaris* thymol.

Seven chemotypes of thyme are known in the western Mediterranian area alone [Bruneton 1995].

Chemistry Of Essential Oils

The total essential oil content of plants is generally very low (<1%). However many therapeutic oils are so potent they are still active in herbal (Galenical) preparations. Upon isolation these oils are highly concentrated, and are widely used in this form by aromatherapists, though rarely for internal consumption. Most oils consist of complex mixtures of chemical compounds, and it is often the unique chemical combination rather than a single component that is responsible for any therapeutic activity. The composition can vary according to the season, time of day, growing conditions and even the genetic make up of the plant. Many oils contain over 50 individual compounds – these can generally be identified using gas chromatography.

Major Categories of Aromatic Oil Compounds

TERPENOIDS

These are constructed from a series of isoprene units linked together in head-to-tail fashion.

They are synthesised via the acetate pathway, as described elsewhere in this book. The most widespread are the **monoterpenes $C_{10}H_{16}$**. These include ketones, aldehydes, alcohols, oxides, phenols, hydrocarbons. Their properties are determined by functional groups – oxygen containing molecules attached to the carbon skeleton [Schnaubelt 1989].

Sesquiterpenes $C_{15}H_{24}$ and **diterpenes $C_{20}H_{32}$** also occur in essential oils. Their properties are less influenced by functional groups.

PHENYLPROPANOIDS

These compounds contain a benzene ring structure with an attached propane (C3) side chain. The most common precursor is cinnamic acid, a derivative of the shikimic acid pathway. They include some aldehydes, phenols and phenolic ethers.

CLASSIFICATION OF ESSENTIAL OIL COMPOUNDS

Compound	Description	Example
Hydrocarbon	Contain only carbon and hydrogen.	Pinene; limonene
Alcohol	Contains a hydroxyl group attached to the terpene structure.	Menthol; terpinen-4-ol; geraniol.
Aldehyde	Terpenoids with a double-bonded O and an H attached to a C.	Citral (geranial/neral); citronellal
Cyclic aldehydes	Aldehyde group attached to a benzene ring	Cinnamic aldehyde; vanillin
Ketone	Contains a carbonyl group C=O	Camphor; thujone
Phenol	Hydroxyl group attached to a benzene ring	Thymol; eugenol; carvacrol
Phenolic ether	Contains an O between C and benzene ring	Safrole; anethole; myristicin
Oxide	Has an O bridging 2 or more carbons	Cineole; ascaridole
Ester	Contains 2 Os attached to a carbon	Methyl salicylate; allyl isothiocyanate

[Cracker 1990]

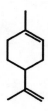

limonene

Hydrocarbons

These are almost universal in monoterpene oils, they are the basic structure from which other terpenoid essential oils are derived. **Limonene**, for example, is the precursor of the main constituents of the monoterpenes in mint – *Mentha* spp.(Lamiaceae), including carvone and menthol [Croteau 1991]. It is also found in citrus oils and dill – *Anethum graveolens* (Apiaceae).

Limonene and other citrus oils have shown great potential as antitumor agents [Tisserand & Balacs 1988/89]. In a study supported by The US National Cancer Institute, limonene and other monoterpenes from dill and caraway oils increased levels of the detoxifying enzymes glutathione S-transferase (GST) in mice tissues, indicating potential cancer chemoprotective activity [Zheng et al 1992]. The authors of this study observe the significance of these compounds as dietary constituents, present in the oils of the citrus family, as well as in dill, celery and other items of human consumption. The hydrocarbons in dill oil are thought to contribute to their diuretic action [Mahran et al 1992]. Limonene and other terpene hydrocarbons have also been found to possess antiviral properties in low concentrations [Schnaubelt 1989].

Pinene

α- & β-**pinene** are widely distributed in plants, with high proportions in **oil of turpentine** (terebinthina) from different species of *Pinus*. Turpentine oil is used as a rubefacient or liniment in rheumatic disease. Pinene has a pleasant aromatic odour and is an important component of many culinary spices, including black pepper – *Piper nigrum* (Piperaceae).

Other monoterpene hydrocarbons include α- & γ-**terpinene**, *p*-**cymene**, **myrcene** and α-**phellandrene**.

Alcohols

Alcohols have a hyroxyl group attached to a C10 hydrocarbon skeleton. Terpene alcohols are so highly valued for their fragrance, healing properties and gentle reaction on skin and membranes they have been termed "friendly molecules" [Schnaubelt 1989]. The names of alcohols always end in "ol" .The most important compounds in this class are **terpinen-4-ol** from tea tree oil and **menthol** from mints (*Mentha* spp).

Alcohols rank with phenols as the most potent antimicrobial essential oil compounds, however they do not contain the irritant properties of the latter.

Tea tree oil

Derived from leaves of *Melaleuca alternifolia* and *M. linariifolia* (Myrtaceae), native to the east coast of New South Wales. To meet the Australian Standard for tea tree oil the terpinen-4-ol content must be at least 30% while that of the oxidised terpene 1,8 cineole must not exceed

menthol

terpinen-4-ol

15% [Williams et al 1988]. The chemotypes of tea tree oil from natural stands depends is genetically determined, so that all commercial plantations are started from seed known to be of the right chemotype. Other terpene constituents of tea tree oil are the alcohol α-**terpineol**, hydrocarbons α- and β-**pinene**, **p-cymene** and γ-**terpinene**, and sesquiterpenes including the unique compound **viridflorene** [Penoel 1990]. Terpinen-4-ol is also a major ingredient in marjoram – *Marjorana hortensis* (Lamiaceae).

While tea-tree oil is not the most powerful antimicrobial available, it is considered by some to be the ideal skin disinfectant due to its activity against a wide range of micro-organisms (both gram+ and gram- bacteria as well as fungi), its low incidence of irritation, and ease of penetration [Altman 1988]. Tea tree oil can be applied to all afflictions of the skin and orifices. The broad antimicrobial activity makes it useful for vaginal irritations since these can result from a variety of pathogenic organisms including yeasts and bacteria [Williams & Home 1995]. However its action is not restricted to that of an antiseptic, it is revered by aromatherapists for its general harmonizing attributes and immune stimulant effects [Penoel 1990].

Peppermint oil

Derived from the dried leaves and flowering tops of *Mentha piperita* (Lamiaceae).

The oil consists of about 50% **menthol**. The taste and odour of peppermint oil are also influenced by some of its minor components, notably the menthol esters **jasmone** and **menthofuran**. The latter compound has a disagreeable odour and is mainly concentrated in young peppermint plants [Samuelsson 1992]. Peppermint is one of the best carminatives and the oil is sometimes administered in capsules for irritable bowel syndrome. Animal studies using peppermint oil demonstrated a significant spasmolytic effect, thought to be linked to the menthol content [Taddei et al 1988]. **Menthol** is an ingredient in several pharmaceutical preparations and inhalants for congestion of the respiratory tract.

Other terpene alcohols

Geraniol, citronellol from rose oil, *Rosa gallica* (Rosaceae) and scented geraniums *Pelargonium* spp. (Geraniaceae). **Nerol** is a steroisomer of geraniol.

Borneol is found in Rosemary oil – *Rosmarinus officinalis* (Lamiaceae)

Linalol from lavender oil – *Lavandula* spp. (Lamiaceae) and ylang ylang – *Cananga odorata* (Annonaceae).

Santalol from heartwood of sandalwood – *Santalum album*, *S. spicatum* (Santalaceae)

Aldehydes

Aldehydes such as those found in citrus oils correspond to their respective alcohol – note that their names end in "al", hence: geraniol, citronellol (alcohols); geranial, citronellal (aldehydes)

Citrus oils

Essential oils are present in leaves, flowers and fruits of plants in the citrus, however the main medicinal oils are found in the fruit peel. The best quality oils come from the bitter orange – *Citrus aurantium* (Rutaceae) and the lemon – *Citrus limon*. Although the hydrocarbon **limonene** is the major constituent, the aroma of the oils is determined by the presence of aldehydes, namely the isomers **geranial** and **neral** – together known as **citral**, and **citronellal**. Citral features as the dominant constituent in other citrus flavoured oils such as those from lemon grass (*Cymbopogon citratus*), lemon balm (*Melissa officinalis*), lemon verbena (*Alloysia triphylla*) and the Australian sweet verbena tree – *Backhousia citriodora*, which consists almost entirely of citral. The lemon scented gum – *Eucalyptus citriodora* (Myrtaceae) – consists mainly of **citronellal** with a small amount of the alcohol citronellol,

geranial

while lemon scented tea tree (*Leptospermum citratum*) contains both citral and citronellal.

Properties of aldehydes

Apart from its pleasant aroma, citral is valued for its sedative, antiviral and antimicrobial properties. Citral rich oils derived from lemongrass and lemon scented tea tree were shown to inhibit *Candida albicans* at more than four times the rate (zone of inhibition) of tea tree oil [Williams & Home 1995]. However many aldehydes are irritants, causing skin sensitivity in some people thereby restricting their use in topical applications. Citral and other aldehydes are also thought to have antitumor properties, though the few clinical trials carried out on them are inconclusive [Tisserand & Balacs 1988/89].

Cyclic aldehydes

These are also known as aromatic aldehydes – they are derived from phenylpropanoids and have no terpene structure. They have characteristically sweet, pleasant odours and are found in some of our most well known herbs and spices such as cinnamon and nutmeg.

benzaldehyde

Cinnamic aldehyde is found in cinnamon and cassia barks – *Cinnamomum* spp. (Lauraceae). **Benzaldehyde** is the main constituent of bitter almond essential oil.

Tisserand and Balacs [1988/89] report on recent research on benzaldehyde, for which anti-tumor properties have been demonstrated in clinical trials. While these aldehydes can also produce skin sensitisation and irritation they are not considered to be toxic [Tisserand 1988].

Phenols

While present in only a relatively few aromatic herbs, phenolic volatile oils are amongst the most potent and potentially toxic compounds found in essential oils. Phenols are represented in both major classes of aromatic compounds – the monoterpenes and phenylpropanoids. The major monoterpene phenols **thymol** and **carvacrol** are found in thyme – *Thymus* spp. and oregano – *Origanum vulgare* (Lamiaceae).

Thymol

Derived from oil of thyme – *Thymus vulgaris* (Lamiaceae). Thyme oil consists entirely of terpenes, the most dominant being a mixture of the phenols **thymol** and **carvacrol**. Other compounds present are the alcohols linalool, geraniol and α-terpineol [Stahl-Biskup 1991].

Thymol is expectorant, antimicrobial, anthelmintic and antispasmodic. It is a dermal and mucous membrane irritant and caution is required in its use. The tincture is a safer means of administration than the oil itself.

thymol

Eugenol

The main source of **eugenol** is clove oil from flower buds of *Syzygium aromaticum* (Myrtaceae), and it is a major constituent of bay leaf and allspice. It has anti-microbial properties akin to thymol, which, coupled with its anaesthetic properties make it an effective disinfectant and cauterising agent in dentistry [Valnet 1980]. Clove oil is an effective topical remedy for toothache and is a powerful stimulant and aromatic. Clove powder is an essential ingredient in the "Composition Powder" made famous by Samuel Thomson. Recent studies demonstrate potent inhibitory effects on lipid peroxidation [Toda et al 1994] and dual inhibition of arachidonic acid and platelet-aggregating factor (PAF) for eugenol [Saeed et al. 1995]

eugenol

Phenols are the strongest antimicrobial agents amongst the monoterpenoid compounds, but potentially very irritant and toxic except in very low concentrations [Schnaubelt 1989].

Phenolic ethers

The majority of these phenols are derived from **phenylpropane**, they are prominent constituents of common spices such as cloves, aniseed, celery seed, basil and tarragon.

The structures are characterised by a side group with two oxygens attached to the phenylpropanoid ring as occurs in **safrole**.

Phenolic ethers when isolated are irritant and toxic, and the oils from which they derive must be used with great care. The spices themselves are generally considered to be safe.

safrole

Safrole from the root bark of the sassafras tree – *Sassafras albidum* and *Camphora* spp., both of the Lauraceae family. It is also found as a minor constituent in cocoa, nutmeg and pepper and was once the principle ingredient of "root beer" [Hall 1973]. Use of this oil is restricted due to suspected carcinogenic properties.

Myristicin from nutmeg and mace – *Myristica fragrans* (Myristcaceae). It also occurs in black pepper, carrot, parsley and dill. Myristicin is structurally related to **safrol** and is toxic in high doses. Nutmeg itself exhibits narcotic and intoxicating properties, though it does have medicinal uses in lower doses [Hall 1973].

Methylchavicol is the main constituent of oil of basil – *Ocimum basilicum* (Lamiaceae). As with other phenols this one is a skin irritant, though it is milder than eugenol which is a component of the French basil varieties [Home & Williams 1990].

Anethol derived from aniseed – *Piminella anisum* (Apiaceae), star anise and fennel.

Apiole derived from parsley seed – *Petroselinum crispum* (Apiaceae).

Ketones
Monoterpenoid ketones are cyclic compounds in which a carbonyl group is attached to a carbon ring. Monocyclic ketones such as **carvone** contain a single carbon ring while dicyclic ketones like **camphor** contain two fused rings.

camphor

Ketones have attained a notoriety because of their alleged toxic and abortifacient properties, however not all ketones are toxic [Tisserand 1988]. Ketones are present in some of the more benign essential oils such as those from spearmint and fennel. On the other hand **thujone**, **pulegone** and related ketones are certainly toxic with potential for inducing convulsions in high doses, and they must never be used during pregnancy. On the positive side ketone volatile oils are mucolytic and of benefit for respiratory congestion [Schnaubelt 1989]. The ketone **carvone** is optically isomeric – one isomer (+)-**carvone** is found in oil of caraway seed – *Carum carvi* (Apiaceae) whereas the other, (-)-**carvone**, is the main constituent of oil of spearmint – *Mentha spicata* (Lamiaceae) [Pavia et al 1982].

Camphor

Derived from the heartwood of the camphor laurel – *Cinnamomum camphora* (Lauraceae).

Much of the camphor used in commerce is prepared synthetically from other monoterpenes. Camphor is a CNS stimulant which is toxic in high doses. It is primarily used as a topical agent for its antipruritic, rubefacient and mucolytic properties.

Thujone

This ketone was originally isolated from the Arbor Vitae – *Thuja occidentalis* (Cupressaceae), however it also occurs in some unrelated plants, particularly those of the *Artemisia genus* (Asteraceae) including wormwood (*A. absinthium*) and mugwort (*A. vulgaris*). Other sources are tansy – *Tanacetum vulgare* (Asteraceae), sage (*Salvia officinalis*) and clary sage (*S. sclarea*) from the Lamiaceae family. Many of these herbs are steeped in folklore with reputations as stimulants, psychedelics, anthelmintics, abortifacients and as ingredients in intoxicating beverages such as vermouth and absinthe [Albert-Puleo 1978]. The above herbs are all classed as emmenagogues and are contra-indicated for pregnancy. Sage oil is the least toxic of this group despite containing 50% thujone, however care is still necessary in its use [Tisserand 1988]. **Pulegone**, a related ketone found in oil of pennyroyal *Mentha pulegium* (Lamiaceae), has similar properties, and the same precautions on its use apply as for thujone containing oils.

Oxides

The oxides are a very small group of essential oils, consisting basically of two unrelated oils, the monoterpene ether **1,8-cineole** and the peroxide **ascaridole**.

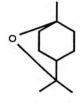

1,8-cineole

Cineole

1,8-cineole is one of the most widely distributed compounds in the plant world, being an oxidized derivative of other monoterpene compounds. Cineole is the major constituent of Eucalyptus oil derived from numerous species of *Eucalyptus* (Myrtaceae) – hence the alternative name **eucalyptol**. It is also the major constituent of oil of cajuput – *Melaleuca cajuputi* (Myrtaceae). Cineole is an expectorant and mucolytic agent, and is a universal ingredient in cough lozenges and other medications.

Eucalyptus oil

Eucalyptus oils vary in aroma and quality according to the level of cineole and of the minor constituents present in each oil. Cineole rich oils (generally from *Eucalyptus globulus*) are preferred for medicinal use where their expectorant property is highly valued, however other species with a different balance of compounds are sometimes preferred – especially by aromatherapists. These include *E. radiata* with the hydrocarbons α-terpineol and pinene and *E. dives* containing phellandrene and the ketone piperitone. All Eucalyptus oils are renowned for their antiseptic qualities, while some species have been shown to inhibit viruses as well. Oils high in cineole and terpene hydrocarbons are considered the most effective against influenza viruses [Schnaubelt 1988/89]. The synergistic activity of using two or more essential oils together can result in an even stronger antimicrobial activity, as demonstrated in an experiment where the addition of basil oil to eucalyptus oil increased the bactericidal activity twenty-fold! [Brud & Gora 1989].

Ascaridole

Ascaridole, a terpene peroxide, is considered to be the most highly toxic of all the essential oils [Tisserand 1988]. It is the main constituent of **wormseed oil** – from *Chenopodium ambrosioides* (Chenopodiaceae) and a lesser constituent of **oil of boldo**, from *Peumus boldo* (Monimaceae). Ascaridole is strongly anthelmintic though its use is limited by its toxic nature. Wormseed is sometimes used in a crude form, ie. powder or extract where it is relatively safe to use in low doses.

ascaridole

Esters

Most esters are formed by reaction of terpene alcohols with acetic acid. They are amongst the most widespread volatile oil compounds, though they are generally present in small amounts. However their distinctively fragrant aromas characterise many of the oils in which they appear. Oil of lavender – *Lavandula* spp. (Lamiaceae) – contains the alcohol **linalol** (also called linalool) along with its ester **linalyl acetate**.

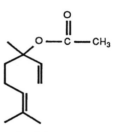

linalyl acetate

Most esters are gentle, non-irritant compounds, whose action is mainly sedative and anti-spasmodic [Schnaubelt 1989]. Examples of these are found in oils of lavender, roman chamomille – *Anthemis nobilis* (Asteraceae), clary sage – *Salvia sclarea* (Lamiaceae) and bergamot – *Citrus aurantium* (Rutaceae). Less benign esters are found in oils of wintergreen and mustard (see below).

Methyl salicylate

Derived from oil of wintergreen – *Gaultheria procumbens* (Ericaceae). **Methyl salicylate** is an aromatic ester derived from cinnamic acid and methanol, though it can now be produced synthetically [Tyler 1988]. It is mainly used in topical applications and linaments as a counter-irritant and antirheumatic. Internal administration is not recommended since it is quite toxic in large doses.

Allyl isothiocyanate

This is a sulphur compound derived from oils of mustard – *Brassica nigra* and spp. and horseradish – *Cochlearia armoracia* (Brassicaceae). These oils are extremely toxic and irritant to the eyes, skin and mucous membranes and should not be used therapeutically – either externally or internally [Tisserand 1988]. Horseradish and mustards are best used as foods or in plant extracts. **Allyl isothiocynate** and similar compounds from the *Allium* genus, are known to have antimicrobial and antitumor properties.

Allyl sulphides

These are a group of sulphur compounds based on the "allyl" group ($CH_2=CHCH_2$) which are found in **garlic** – *Allium sativa*, **onion** – *A. cepa*, and other members of the *Allium* genus (Liliaceae). Their common precursor is the sulphur-containing amino acid **cysteine**. The complex chemistry of garlic and onions is best reviewed by Eric Block [1985] and more recently by Sendl [1995]. Extraction techniques using different solvents has assisted in identification of oxidation and decomposition products of garlic and onions. These include:

  diallyl disulphide – product of steam distillation of garlic
  allicin – the oxide of diallyl disulphide, and source of garlic odour
  ajoene – formed from decomposition of allicin
  lacrimatory factor – converted from precursor by enzymic action when slicing onion [Block 1995].

allicin

(E)-ajoene

Apart from antimicrobial properties noted above, allyl derivatives such as **ajoene** have demonstrated activity against platelet aggregation, inhibition of proinflammatory prostaglandins, antitumor and hypoglycaemic effects [Reuter 1995].

RESINS

Resins are solid, brittle substances secreted by plants into special ducts, often as a response to damage to the plant by wounding, wind damage etc. Their main role appears to be protection of the plant from attack by fungi and insects. Resins are difficult to classify because of their amorphous nature, they are complex mixtures which include lignans, resin acids, resin alcohols, resinotannols, esters and resenes [Tyler 1988].

Resins are insoluble in water but soluble in alcohol, fixed oils, volatile oils. They are heavier than water and volatile oils, and have high boiling points. They are translucent and burn with a characteristic smoky flame – hence their use in incense. They have fixative actions hence their use in crafts and industry.

While classifying individual resins is difficult, they are sometimes classified as mixture with other plant constituents eg. gum-resins, oleo-gum-resins, glycoresins. One of the most well known resins comes from the *Pinus* genus (Pinaceae) and is known as **rosin**. This amber coloured resin is mainly used in varnishes and other industrial products.

Balsams

These are resinous mixtures that contain cinnamic and/or benzoic acid or their esters.

Benzoin

This balsamic resin is derived from the bark of *Styrax benzoin* (Styraceae) trees in S.E. Asia. **Benzoin** contains cinnamic, benzoic and triterpene acids. Its action is anti-septic, stimulant, expectorant, diuretic and antifungal. It is used as a food preservative as well as an ingredient in pharmaceutical preparations such as Whitfield's Ointment (with salycilic acid) for ringworm, athletes foot [Tyler 1988].

Podophyllin

Derived from dried rhizomes and roots of *Podophyllum peltatum* (Berberidaceae), a plant which originates in the forests of central and eastern USA.

Contains 3.5-6% **podophyllin** resin consisting of the lignans:

> **podophyllotoxin** 20%
> **a & b peltatin** 15% – purgative principle
> + lignan glycosides (lost during preparation of resins)

Podophyllin powder has a peculiar bitter taste and is highly irritating to mucous membranes, especially of the eyes. Its caustic nature is utilized in the form of topical applications for warts and condylomas. Internally it acts as a drastic though slow acting purgative. It is also anti-mitotic, ie. stops cell divisions and is sometimes used in leukemia, but problems with side effects limits its use.

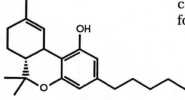

kavain

Kava Kava – *Piper methysticum* (Piperaceae)

This large shrub is widely cultivated in Oceania. The dried rhizome and root are used in preparation of an intoxicating beverage. Kava contains 5-10% resin, made up of lactones known as kava pyrones – **kavain** $C_{14}H_{14}O_3$; **methysticin** $C_{14}H_{15}O_3$.

Kava pyrones are potent, centrally acting skeletal muscle relaxants. The action is hypnotic; antipyretic; sedative; local anaesthetic; smoothe muscle relaxant; antifungal. No interaction with benzodiazepine drugs or with moderate consumption of alcohol occurs, nor does kava impair mental alertness. However continually chewing the root can destroy tooth enamel, and eventually becomes habit-forming [Bone 1994].

tetrahydrocannabinol

Cannabis

Cannabis resin or Indian hemp is derived from dried flowering tops of pistillate plants of *Cannabis sativa* and *C. indica* (Cannabaceae). The plants contain 15-20%

resin consisting mainly of Δ9-**tetrahydrocannabinol** or THC. THC is technically a benzotetrahydropyran and its structure derives from both phenolic and terpenic precursors [Bruneton 1995].

THC is a lipophilic compound which is quickly absorbed but slowly excreted. It is responsible for most of the pharmacologic actions of Cannabis. Therapeutically it is euphoric, sedative, anti-nauseant, appetite stimulant, bronchodilator and reduces intra-ocular pressure in glaucoma. However legal restrictions mean that most use of the herb is illicit. The euphoric effects of the herb are well documented, as are the psychological dependence and disruptive effects on memory and cognitive processes. Its use has also been linked to infertility [Sethi 1991]. However acute toxicity is very low and there are no documented cases of fatalities from Cannabis use [Bruneton 1995].

Major resin & oleo-gum-resin containing herbs

Myrrh
Oleo-gum-resin exudes from incisions made in bark of myrrh trees in North Africa and the Arabian Peninsula – *Commiphora molmol* (Burseraceae) – and India (*C. mukul*), where it is known as guggul. The main constituents are resin 25-40%, gum 60%, volatile oil 2.5-8% along with a bitter principle. Its actions are anti-septic, antimicrobial, astringent and stimulant.

Capsicum
Capsaicin is the pungent principle derived from fruits of cayenne pepper – *Capsicum annum, C. frutescens* (Solanaceae). Capsaicin is a phenolic amide with the formula $C_{18}H_{27}NO_3$. It is present in the fruit at a level of only 0.02% yet its taste is detectable even in minute doses. The compound acts as a local anaesthetic and pain reliever through a complex mechanism [Palevitch and Cracker 1993].

Ginger

The oleo-resin from the rhizome of ginger – *Zingiber officinalis* (Zingiberaceae) contains phenolic arylalkanones which are related to the pigment **curcumin**, from turmeric. These are known as **gingerols**, and are the compounds responsible for the familiar pungent taste of ginger. Gingerols, derived from phenylpropanes, have cholagogue and hepatoprotective effects in rats [Bruneton 1995]. Recently, related compounds (diarylheptenones) with antifungal properties were identified in ginger and named **gingerenone A, B & C** [Endo et al. 1990].

Asafoetida – *Ferula assa-foetida* (Apiaceae)

Asafoetida is one of the most pungent of all spices, it is obtained from the rhizomes and roots of a shrub that is native to the south-west Asian region. It is rich in an oleo-gum-resin, made up of 6-17% volatile oils, 40-64% resin and 7-25% gum. The volatile oil contains disulphide compounds similar to those in garlic. The resin consists of **farnesiferols** which are sesquiterpenoid coumarins, along with ferulic acid and asaresinatannols [Bradley 1992].

FIXED OILS

Unlike essential oils, fixed oils are non-volatile. They are classified as primary metabolites, and as nutritional substances they are not covered here in any detail. However some of the fixed oils have pharmacological properties which are of interest to the medical herbalist. Fixed oils are composed of fatty acids, hydrocarbon chains with a methyl (CH_3) group at one end (Ω end), and a carboxyl group (COOH) at the other (δ end).

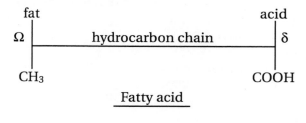

Fatty acid

Carbon bonding

Fatty acids are classified according to their number of double bonds and to which end of the carbon chain the double bonds are nearest to.

1. saturated fatty acids – no double bonds eg. **stearic acid**
2. monounsaturated fatty acids – one double bond eg. **oleic acid**
3. polyunsaturated fatty acids – two or more double bonds eg. **linoleic acid**

Essential fatty acids

Essential fatty acids (EFAs) are present in all the bodies cells, and are especially concentrated in the brain and central nervous system. The body is unable to synthesize these acids hence they must be provided in the daily diet. Linolenic acid is one of the most important EFAs. All EFAs are in the Ω 3 and 6 series fatty acids.

Saw Palmetto – *Serenoa repens* syn. *Sabal serrulata* (Palmae)

Fruits obtained from this species of palm are very rich in fats, including **oleic**, **lauric**, **myristic** and **linoleic** fatty acids. Lipophilic extracts of the herb inhibit the enzyme (5α-reductase) responsible for converting the male hormone testosterone to 5α-dihydrotestosterone, the metabolite associated with prostate enlargement. Clinical trials support the use of saw palmetto extracts in the treatment of benign prostatic hypertrophy (BPH) while its anti-androgenic properties may also assist in acne, female hirsutism and even baldness. Recent studies showed the inhibition of 5α-reductase corresponds with degree of saturation of the fatty acids. Saturated fatty acids show no inhibition, mono-unsaturated fatty acids showed slight inhibition while the unsaturated acids (linoleic, linolenic) were the strongest

oleic acid

inhibitors [Niederprum et al. 1994]. The length of the carbon chain also influenced the degree of inhibition.

Evening Primrose Oil

The oil is derived from seed of *Oenothera biennis* (Onagraceae). Evening primrose oil is rich in the Ω 6 fatty acids, **linoleic** and γ-**linolenic** acids. While linoleic acid is easily obtained as a dietary oil, γ-linolenic acid (GLA) is relatively rare in foods and is usually converted to it in the body from dietary linoleic acid. However there are many factors that may be responsible for blocking the enzymic process responsible for this, and atopic individuals with inherent susceptibility to allergic disorders, excema and asthma included, are usually deficient in GLA [Willard 1992]. Therapeutically, evening primrose oil and other preparations containing GLA are used for atopic excema, PMS and mastalgia, rheumatoid arthritis, endometriosis and schizophrenia [Horrobin 1990]. GLA is also present in oil of borage (*Borago officinalis*) and other plants in the Boraginaceae.

EPAs

This is an Ω 3 fatty acid with the long name of **eicosapentaenoic acid** (EPA). EPAs are mainly found in fish oils, but are present in certain vegetable oils such as linseed, canola and walnut oils. This long chain (20-carbon) polyunsaturated fatty acid inhibits pro-inflammatory prostaglandins including leukotriene – responsible for bronchospasms that occur in asthma. High levels of dietary EPAs reduce thrombosis, by activation of a blood clotting inhibitor prostaglandin and partial inactivation of the platelet aggregation factor. The resulting decrease in blood viscosity leads to prolonged bleeding time and should be taken with caution by those on antithrombotic thaerpy.

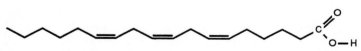

γ- linolenic acid

Other effects recorded are decreases in cholesterol, VLDL and triglyceride levels – indicating that EPAs are generally protective against diseases of the vascular system. [Marderosian & Liberti 1988].

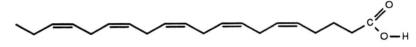

eicosapentaenoic acid (EPA)

References

Albert-Puleo, M. 1978. Mythobotany, pharmacology, and chemistry of thujone-containing plants and derivatives. *Economic Botany*.

Altman, P. 1988. Australian tea tree oil. *Current Drug Information 2*: 62-64.

Block, E. 1985. The Chemistry Of Garlic And Onions. *Scientific American* 252: 94-99.

Bone, K. 1994. Kava – a safe herbal treatment for anxiety Part 2. *Mediherb Professional Newsletter* 39.

Brud, W. & Gora, J. 1989. *Biological activity of essential oils and its possible applications*. Paper presented at 11th International Congress of Essential Oils, Fragrances and Flavours, India.

Cracker, L. 1990. Herbs & Volatile Oils. *The Herb, Spice and Medicinal Plant Digest* 8: 1-5

Croteau, R. 1986. Biochemistry of Monoterpenes and Sesquiterpenes of the Essential Oils. In Craker & Simon (eds) *Herbs, Spices & Medicinal Plants Vol. 1* Oryx Press, USA.

Croteau, R. 1991. Metabolism of monoterpenes in mint (*Mentha*) species. *Planta Medica* 57: S10-S14.

Endo, K. Kanno, E. & Oshima, Y. 1990. Structures of antifungal diarylheptenones, gingerenones, isolated from the rhizomes of *Zingiber officinale*. *Phytochemistry* 29: 797-799.

Guba, R. 1992. *The Fragrant Pharmacy – An Introduction to Aromatic Medicine*. Centre for Aromatic Medicine, Melbourne

Hall, R. 1973. *Toxicants occurring naturally in foods*. National Academy of Sciences, Washington D.C.

Home, V. & Williams, L. 1990. The diversity of basil oils. In *Modern Phytotherapy – the clinical significance of tea tree oil and other essential oils Vol. 3*. Macquarie University, Sydney.

Horrobin, D. 1990. Gamma linolenic acid: an intermediate in essential fatty acid metabolism with potential as an ethical pharmaceutical and as a food. *Rev. Contemp. Pharmacother.* 1: 1-45.

Mahran, G. Kadry, H. Thabet, C. et al. 1992. GC/MS Analysis of volatile oil of fruits of *Anethum graveolens. Int. J. Pharmacognosy* 30: 139-144

Marderosian, A. & Liberti, L. 1988. *Natural Product Medicine. A scientific guide to foods, drugs, cosmetics.* George F. Stickley, Philadelphia.

Niederprum, H. Schweikert, H. & Zanker, K. 1994. Testosterone 5α-reductase inhibition by free fatty acids from *Sabal serrulata* fruits. *Phytomedicine* 1: 127-133.

Palevitch, D. & Cracker, L. 1993. Nutritional and medical importance of red peppers. *The Herb, Spice, & Medicinal Plant Digest* 11: 1-4.

Pavia, D. Lampman, G. & Kriz, G. 1982. *Introduction to organic laboratory techniques.* Saunders College Publishing, U.S.A.

Peneol, D. 1990. The place of the essential oil of *Melaleuca alternifolia* in aromatic medicine. In *Modern Phytotherapy – the clinical significance of tea tree oil and other essential oils Vol. 3.* Macquarie University, Sydney.

Reuter, H. 1995. *Allium sativum* and *Allium ursinum*: Part 2, Pharmacology & medicinal application. *Phytomedicine* 2: 73-91.

Saeed, S. Simjee, R. Shamim, G. & Gilani, A. 1995. Eugenol: a dual inhibitor of platelet-activating factor and arachidonic acid metabolism. *Phytomedicine* 2: 23-28.

Samuelsson, G. 1992. *Drugs of Natural Origin.* Swedish Pharmaceutical Press.

Schnaubelt, K. 1988/89. Potential application of essential oils in viral diseases. *Int. J. Aromatherapy* 1/2: 32-35.

Schnaubelt, K. 1989. Friendly Molecules. *Int. J. Aromatherapy* 2: 20-22

Sendl, A. 1995. *Allium sativum* and *Allium ursinum*: Part 1, Chemistry, analysis, history, botany. *Phytomedicine* 4: 323-339.

Sethi, N. Nath, D. Singh, R. & Srivastava, R. 1991. Antifertility and teratogenic activity of *Cannabis sativa* in rats. *Fitoterapia* LXII: 69-71.

Stahl-Biskup, E. 1991. The chemical composition of *Thymus* oils. *J. Ess. Oil. Res.* 3: 61-82.

Taddei, I. Giachetti, D. Taddei, E. & Mantovani, P. [1988]. Spasmolytic activity of peppermint, sage, and rosemary essences and their major constituents. *Fitoterapia* LIX: 463-468.

Tisserand, R. 1988. *The Essential oil safety data manual.* Tisserand Aromatherapy Institute, U.K.

Tisserand, R. & Balacs, T. 1988/89. Essential oil therapy for Cancer. *Int. J. Aromatherapy* 1/2: 20-25.

Toda, S. Ohnishi, M. Kimura, M. & Toda, T. 1994. Inhibitory effects of eugenol and related compounds on lipid peroxidation induced by reactive oxygen. *Planta Medica* 60: 282.

Valnet, J. 1980. *The Practice of Aromatherapy.* Destiny Books, New York.

Willard, T. 1992. *Textbook of Advanced Herbology.* Wild Rose College of Natural Healing, Canada.

Williams, L. & Home, V. 1995. A comparative study of some essential oils for potential use in topical applications for the treatment of the yeast *Candida albicans. Aust. J. Med. Herbalism* 7: 57-62.

Williams, L. Home, V. Zhang, X. & Stevenson, I. 1988. The composition and bactericidal activity of oil of *Melaleuca alternifolia* (Tea tree oil). *Int. J. Aromatherapy* 1: 15-17

Wrigley J.& Fagg M. 1990. *Aromatic Plants.* Angus & Robertson, Sydney.

Zheng, G. Kenny, P. & Lam, L. 1992. Anethofuran, carvone, and limonene: Potential cancer chemoprotective agents from dill weed oil and caraway oil. *Planta Medica* : 338-341.

Chapter 7
ALKALOIDS

Typical alkaloids are derived from plant sources, they are basic, contain one or more nitrogen atoms (usually in a heterocyclic ring) and they have marked physiological effects on humans or animals.

Willards Definition adds the following conditions:

Molecule must contain nitrogen connected to at least 2 carbon atoms, and have at least 1 ring. The compound cannot be a structural unit of macromolecular cellular substances, and cannot serve as a vitamin or hormone. [Willard 1982: 289]

Most alkaloids are derived at least partly from various amino acids as their direct precursors.

Discovery

The isolation of morphine by Seturner in 1806 led to the discovery of several more alkaloids over the next 15 years, including emetine, piperine, caffeine and quinine. The term alkaloid was first applied by Meissner, a German pharmacist, and originally referred to all plant alkalis.

Natural Occurrence

Alkaloids are found in 15-30% of all flowering plants, and are particularly common in certain familes such as Fabaceae; Liliaceae; Ranunculaceae; Apocynaceae; Solanaceae and Papaveraceae. The most widely occurring alkaloids are caffeine and berberine.

While the higher plants are the major source of alkaloids, they are also known to occur in lower plants such as horsetails and algae, in fungi and other microorganisms, insects and even the organs of mammals [Kapoor 1995]. In plants they may be found in roots, rhizomes, leaves, bark, fruit or seeds. Over 40 alkaloids may occur in a single plant eg. *Vinca major*.

PROPERTIES OF ALKALOIDS

Alkaloids are generally white or colourless crystalline solids containing oxygen. A minority of alkaloids are oxygen free, existing as liquids. These include **nicotine, coniine, sparteine** and **lobeline**. Very few coloured alkaloids exist, exceptions are **sanguinarine** (red) and **chelidonine** (yellow).

With few exceptions (**colchicine**) alkaloids are all alkaline, turning red litmus paper blue.

Many are precipitated by various reagents eg. Meyers (mercuric chloride & potassium iodide). Many give more or less characteristic colour reaction with certain reagents. Most are susceptible to destruction by heat, some by exposure to air and light.

Solubility

Most alkaloids are practically water insoluble. Water soluble alkaloids include **ephedrine** and **colchicine**. They are soluble in organic solvents – chloroform, ether, alcohol. Alkaloidal salts are generally soluble in water and alcohol.

Nomenclature

All alkaloid names end in "ine". Otherwise their names have a number of origins:

1. From generic name of plant – hydrastine
2. From specific name of plant – cocaine
3. From common name of plant – ergotamine
4. Physiological activity – emetine
5. Discover's name – lobeline

Where more than one similar alkaloid are present in a plant, prefixes and suffixes may be added to eg. **quinidine, hydroquinine** are present along with **quinine** in *Cinchona* spp.

PHARMACOLOGICAL ACTIONS

Plant alkaloids usually have profound physiological actions in humans with nervous system effects being the most prominent. There are several good reviews published on this subject [eg. Robinson 1986, 1981; Marini Bettolo 1986]. Examples of some of the more dramatic actions of alkaloids are:

> analgesics/ narcotics – morphine
> mydriatics – atropine
> miotics – pilocarpine
> hypertensives – ephedrine
> hypotensives – reserpine
> bronchodilator – lobeline
> stimulants – strychnine
> antimicrobials – berberine
> antileukemic – vinblastine

CLASSIFICATION OF ALKALOIDS

Alkaloids are a large and diverse group of chemical compounds that defy easy classification.

They are commonly grouped together according to their ring structures. Two major divisions can be made:

> Heterocyclic alkaloids – regarded as most typical
> Non-heterocyclic alkaloids – also known as protoalkaloids or biological amines eg. ephedrine.

Major Alkaloidal Groups and examples

Pyridine/ piperidine – nicotine
Tropane – atropine
Quinoline – quinine
Isoquinoline – berberine
Quinolizidine – sparteine
Pyrrolizidine – senecionine
Indole – reserpine
Imidiazole – pilocarpine
Alkaloidal amines – colchicine
Purine alkaloids – caffeine

PYRIDINE – PIPERIDINE ALKALOIDS

These alkaloids have their N atoms in typical 6-membered rings, which in the case of pyridine is a benzene ring. The precursors are generally ornithine and nicotinic acid though lobeline has a unique biosynthesis (see below).

Main Alkaloids

Coniine (α-propylpiperidine) is found in spotted hemlock – *Conium maculatum* (Apiaceae). It is a piperidine structure with a short aliphatic side chain, and is a volatile oily compound. Coniine is very toxic and causes death by paralysis.

piperidine

Nicotine (1-menthyl-2 (3-pyridyl) pyrrolidine) [$C_{10} H_{14} N_2$]

Found in tobacco derived from *Nicotiana tabacum* and other plants of the Solanaceae, including the Australian pituri (*Duboisia hopwoodii*) whose properties were exploited by Aborigines in the central desert [Watson et al 1983]. Nicotine also occurs in many remotely related plants such as club mosses. This is possible since the precursor nicotinic acid (vitamin B3) is of widespread distribution in the plant family. Pure nicotine is a colourless oily alkaloid, while salts of nicotine are readily water soluble. The prime pharmacological alkaloid in tobacco is **L-nicotine** (0.5-10%), along with **nornicotine**, **anabasine** and **nicotyrine**. Structurally nicotine consists of a simple linking of pyridine and pyrollidine rings.

nicotine

Piperine (1-piperoylpiperidine)

Found in *Piper* spp. – *Piper nigrum* (black pepper); *Piper longum* (long pepper). Both peppers are of major importance in Ayurvedic medicine. **Piperine** has been shown to have hepatoprotective effects though less potent than silymarin [Koul & Kapil 1993].

Arecoline (arecaidine methyl ester)

Found in betel nut derived from the palm *Areca catechu*, widely used in many countries as a masticant. *Areca* also contains about 15% tannins.

Lobeline

Found in *Lobelia inflata* along with **lobelanine** and **lobelanidine**. The main ring in lobeline is derived from lysine via piperidine, while the 2 benzene rings it contains derive from phenylalanine via the shikimic-acid pathway [Samuelsson 1992].

lobeline

Lobelia is a relaxant and bronchodilator originally made famous by Thomson and the Physiomedical school of herbalists, however its use is now mainly restricted to medical practitioners.

QUINOLINE ALKALOIDS

This is an example of a bicyclic ring system with the fusion of a benzene and a pyridine ring. Biosynthetically they are related to indole alkaloids since both groups are derived from the same 2 precursors, tryptophan and loganin – a monoterpene iridoid [Samuelsson 1992].

Main Alkaloids

Quinine (6-methoxycinchonine) $C_{20} H_{24} N_2 O_2$

Quinine is found in Peruvian bark – *Cinchona* spp. (Rubiaceae), a tree which originates in the Andes mountains. The major species used are *Cinchona succirubra* and *C. ledgeriana*. **Quinidine** is the isomer of quinine.

quinoline

Cinchona bark contains 25 closely related alkaloids, including the main therapeutic alkaloids in this category. The average yield is 6-7% (25% quinine, 5% quinidine).

The bark also contains cinchotannic acid (2-4%) which yields insoluble 'cinchona red'.

Uses of quinine:
- anti-malarial
- skeletal muscle relaxant (treatment of nocturnal night cramps)
- febrifuge for colds, influenza etc.
- tonic water – quinine is very bitter.

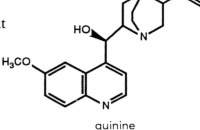

quinine

Uses of quinidine:
- cardiac arrythmias
- atrial flutter/ fibrillation

Quinidine depresses myocardial excitability, conduction, velocity and contractility. It is used in the form of quinidine sulphate.

Side effects of quinine drugs:
- Tinnitus ('cinchonism'); skin rash; vertigo; unusual bleeding/bruising; visual disturbances.

Other alkaloids

Quinoline alkaloids are particularly common amongst plants of the Rutaceae family.

Ruta graveolens or common rue contains 30 known alkaloids of the quinoline type [Harborne & Baxter 1993]. These include **arborinine** and **γ-fagarine**.

Another Rutaceous plant with a similar alkaloidal spectrum to rue is prickly ash bark – *Xanthoxylum* spp. Two new quinoline alkaloids isolated from *Xanthoxylum simulans* were shown to have cytotoxic and antiplatelet activities [Chen et al. 1994].

ISOQUINOLINE ALKALOIDS

These alkaloids result from the condensation of a phenylethylamine derivative with a phenylacetaldehyde derivative. Both moieties are derived from the same precursors. The precursors are phenylalanine or tyrosine.

isoquinoline

Isoquinoline alkaloids are most frequently found in the Papaveraceae, Berberidaceae and Ranunculaceae families. This is a very large class of medicinally active alkaloids,

papaverine

which for convenience can be divided into the following sub-classifications:

Morphinane alkaloids – morphine, codeine, thebaine (phenanthrene derivatives)

Benzylisoquinolines – papaverine, oxyacanthine, turbocurarine

Protoberberines – berberine, hydrastine, palmatine.

Protopines – cryptopine, protopine

Benzophenanthradine – chelidonine, sanguinarine.

Ipecac alkaloids – emetine, cephaeline

Aporphine alkaloids – boldine

PROPERTIES OF ISOQUINOLINE ALKALOIDS

Reported pharmacological properties include the following:

antispasmodic (papaverine) [Martin et al.1993]

antimicrobial, antitumor [Wu et al.1989]

antifungal (jatrorrhizine) [Grayer & Harborne 1994]

antiinflammatory, cholagogue & hepatoprotective, antiviral, amoebicidal [De Silva 1983]

antioxidant (oxyacanthine) [Muller & Ziereis 1994]

enzyme inhibitors [Robinson 1986]

Opium poppy – *Papaver somniferum* (Papaveraceae)
Opium is the dried latex obtained from incisions made in unripe fruits of the opium poppy. The official opium drug is standardized to contain 10% morphine. Opium contains over 40 alkaloids which are usually combined with meconic acid, a signature compound for identifying opium. Some major opium alkaloids are listed below [Kapoor 1995]:

morphine [8-14%] narcotic analgesic & hypnotic

noscapine [4-8%]antitussive; no narcosis

codeine [2.5%-3.5%] antitussive; mild narcotic and analgesic

papaverine [0.5-1%] smoothe muscle relaxant, antitussive

thebaine [0.1-2%] alternative source for synthesis of codeine

Heroin is diacetylmorphine – a synthetic derivative which is more toxic and habit-forming than morphine.

Morphine $C_{17} H_{19} NO_3$

Morphine is a complex phenolic compound whose pentacyclic structure is derived from tyrosine. It is analgesic, narcotic and a powerful respiratory depressant which was previously used in cough elixers. Morphine induces euphoria and dependency in some people, anxiety and nausea in others. The CNS effects occur through stimulation of specific receptors. Opiate receptors are widely distributed in animals, they respond to both endogenous transmitters (peptides) and ingested plant alkaloids. The main receptor types are:

δ emotional; λ sedative; μ analgesic; σ psychotomimetic [Robinson 1986.]

morphine

Other effects of morphine include decrease in pupilliary size, reduced HCL secretion in the stomach and constipation. Overdose of morphine can cause death through respiratory arrest [Kapoor 1995].

Ipecac – *Cephaelis ipecacuahna* (Rubiaceae)

Dried rhizome and roots are obtained from low bushes indigenous to Brazil. It contains 5 isoquinoline alkaloids, including **emetine, cephaline, psychotrine**. Emetine hydrochloride is used in medicine as an antiprotozoan. Syrup of ipecac is an emetic and poison antidote.

Emetine is a GIT irritant causing reflex increase in respiratory secretions. It acts as emetic in higher doses, and is indicated in chronic bronchitis and whooping cough.

emetine

Curare – a muscle relaxant drug which was originally used as an arrow poison by Amazonian Indians. The traditional curare is prepared by a secret recipe thought to involve a number of plants [Plotkin 1993]. Plant sources of curare include *Strychnos castelnaei* and spp. in the Loganacece family and *Chondodendron tomentosum* in the

Menispermaceae family. **Tubocurarine**, a benzylisoquinoline dimer, is the major alkaloid in the curare plants. It exhibits paralysing effects on skeletal muscles, and is used as a muscle relaxant in surgical procedures. It controls convulsions caused by the toxic alkaloid strychnine.

Chelidonine – a yellow alkaloid derived from *Chelidonium majus*. It has an analgesic action similar though milder than morphine – its action lasts from 4-48 hours [Huang 1993].

Berberine – a protoberberine alkaloid whose salts form yellow crystals. It is found along with related alkaloids in: *Hydrastis canadensis* **golden seal** (Ranunculaceae); *Berberis* spp. (Berberidaceae): *B. vulgaris* **common barberry**; *B. aquifolium* or *Mahonia aquifolium* **oregon grape root**; *B. aristata* **Indian barberry**.

Actions of berberine include amoebicidal; antibacterial; antifungal; cholagogue; hepatoprotective; tyramine inhibitor; elastase inhibitor – helps repair tissues and reduce inflammation; antitumor [Pizzorno & Murray 1986]. Berberine has been shown to have a negative inotropic effect on the heart, it markedly reduces atrial rate. It also has an antiarrhythmic action [Huang 1993].

Boldine ([S]-2,9-dihydroxy-1,10-dimethoxyaporphine) Boldine is found in the leaves and bark of *Peumus boldo* (Monimiaceae), an evergreen tree native to Chile. Boldine imparts choleretic, cholagogue, antioxidant and smoothe muscle relaxant properties to the herb [Speisky et al 1991] which is used primarily in the treatment of gallstones.

TROPANE ALKALOIDS

These are complex molecules containing pyrrolidine and piperidine ring structures, derived from precursors ornithine and phenylalanine. Isoleucine and acetate also play a roll in the biosynthesis of tropane structures. Alkaloids used in medicine are restricted largely to the

Solanaceae family, apart from cocaine which comes from the coca plant in the Erythroxylaceae family. Esterification of the nitrogenous moeity with tropic acid is a structural type thought to be unique to the Solanaceae family [Roddick 1991; Evans 1990]. The major alkaloids are **hyoscamine** and **hyoscine** (= **scopalamine**). **Atropine**, the racemic form of hyoscamine generally occurs in trace amounts only.

L-hyoscyamine

These alkaloids and their near relatives are found in varying concentrations in 22 genera of the Solanaceae. Some of the best sources are:

Atropa belladonna – deadly nightshade
Datura stramonium – thornapple
Hyoscyamus niger – henbane
Mandragora officinarum – mandrake
Duboisia myoporoides, D. leichhardtii – corkwood

Pharmacological actions of hyoscamine and atropine

Hyoscamine acts on tissue cells innervated by post-ganglionic cholinergic fibres of the parasympathetic nervous system – it is anti-muscarinic and a parasympathetic depressant.

It has a spasmolytic effect on bronchial and intestinal smoothe muscles. The mydriatic effect (inhibits contraction of iris muscle) is more pronounced than that of atropine. It reduces salivary and sweat gland secretions, controls excess motor activity in GIT → antidiarrhoeal effect, and reduces rigidity and tremors in Parkinsonism. Scopalamine is a CNS depressent, used for motion sickness "patches", whereas hyoscamine and atropine are CNS stimulants [Roddick 1991]. Atropine has been used as an antidote in cases of poisoning by cholinesterase inhibitors eg. phytostigmine; organophosphates.

Toxicity of Tropane Solanaceous Alkaloids

Dilated pupils, impaired vision, dryness of skin and secretions, extreme thirst, hallucinations, loss of conciousness.

Datura

Datura stramonium and other species. Also known as stramonium or thornapple.

Contains over 30 tropane alkaloids mainly hyoscamine and hyoscine, as well as N-oxides of hyoscamine and hyoscine, tigloylmeteloidine and nicotine. Indicated for asthma, pertussis, muscular spasm & excess salivation in parkinsonism. [Bradley 1992].

Stramonium is listed in Schedule 1 – dangerous poisons and 2 – medicinal poisons (containing 0.25% or less of alkaloids calculated as hyoscamine) of the NSW Poisons List. Exceptions are made in preparations for smoking or burning.

Duboisia

Duboisia myoporoides; D. leichhardtii. Also known as corkwood.

Unlike the pituri (*D. hopwoodii*), the nicotine containing species that inhabits the central and western regions of Australia, these two tree/shrubs are restricted in their distribution to the east coast and adjacent area. Corkwood leaves contain the highest levels of atropine and tropane alkaloids in the world, and since WW2 have replaced *Atropa belladonna* and *Hyoscymus niger* as the leading source of these alkaloids [Barnard 1952; Roddick 1991].

D. myoporoides exists naturally in distinct chemical races, some of which have higher levels of tropane alkaloids. In commercial cultivation a hybrid form of the species has been utilized whose leaves contain up to 7% alkaloids [Roddick 1991; Evans 1990].

Cocaine

Erythroxylon coca (Erythroxylaceae)

Coca leaves are obtained from shrubs/small trees indigenous to South America.

Alkaloids present include cocaine, cinnamylcocaine, tropacocaine, valerine.

Cocaine is the methyl ester of benzoylecgonine, and can be semisynthetically produced from ecgonine [Tyler 1988].

(-)-cocaine

Pharmacology of cocaine

Cocaine is relatively unstable and is more often administered as cocaine hydrochloride which is a more stable form. Cocaine blocks nerve conduction upon local application, hence its employment as an anaesthetic (eg. Novacaine). In large doses it is a cerebral stimulant and narcotic. Its adrenergic action is due to inhibition of reuptake of noradrenaline, creating an amphetamine-like effect, though only for very short duration.

Toxicology of cocaine

Cocaine is quickly absorbed into the lungs, heart and brain with almost instant effects, leading to psychic dependence and tolerance [Tyler 1988]. It causes destruction of mucous membranes such as those of the nose. Coca leaf is chewed by natives, and appears to be relatively harmless in this form.

QUINOLIZIDINE ALKALOIDS

These are also referred to as 'lupin' alkaloids since they were first discovered in *Lupinus* spp. They occur primarily in the Papillionaceae family. The quinolizidine structure consists of two carbon rings with a shared nitrogen atom. Their precursor is lysine.

The major quinolizidine alkaloids are **sparteine** found in broome and greater celandine (*Chelidonium majus*), and **cytisine** from lupins. **Myrine** comes from bilberry (*Vaccinium myrtillis*) in the Ericaceae family.

quinolizidine

Sparteine from *Sarothamnus scoparius* (Scotch broom).

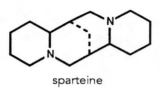

sparteine

Sparteine is a tetracyclic, oxygen free alkaloid. It is oxytocic, cardiac stimulant and diuretic. It binds strongly to muscarinic receptors and (less strongly) to nicotinic receptors [Schmeller et al 1994]. It is a peripheral vasoconstrictor.

Sparteine has a similar cardiac action to quinidine [Samuelsson 1992]. The action of broom in this regard is modified by the presence of flavonoids. It has an ergot-like action on the uterus and has been used as a substitute for ergot drugs. Some sources claim broom is narcotic and it has been used in smoking mixtures, though this practice is considered dangerous.

PYRROLIZIDINE ALKALOIDS

pyrrolizidine

These alkaloids contain two fused 5-membered rings in which a N atom is common to both rings. A hydrogen atom usually occurs opposite the N atom in the α-position, with a hydroxymethyl substitute at the adjacent C-1 position. The precursor is ornithine.

These alkaloids are widespread in the Boraginaceae, Asteraceae and Papillionaceae families, the majority occurring in the mega-genus *Senecio*, all 1500 species of which are thought to contain them, including the common ragwort (*S. jacobeae*) [Harborne & Baxter 1993].

PAs have been associated with veno-occlusive disease and herbs which contain them are scheduled and unavailable to herbalists. This includes comfrey (*Symphytum* spp.); borage (*Borago officinalis*); coltsfoot (*Tussilago farfara*); liferoot (*Senecio aureus*); lungwort (*Pulmonaria officinalis*). Alkaloids include **symphytine, echimidine**.

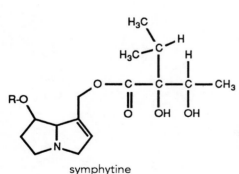

symphytine

The alkaloids also occur as water soluble N-oxides.

Few therapeutic effects have been postulated for pyrrolizidine alkaloids. The main interest has been in their toxicity which is dealt with in numerous other publications [eg. Bruneton 1995]

INDOLE ALKALOIDS

A very large group of alkaloids whose basic structure contains a pyrrole ring fused to a benzene ring. Biosynthesis of pure indoles involves the amino acid tryptophan as precursor, while that of a major sub-group of the indoles, including the *Catharanthus* and *Rauvalfia* alkaloids, have a second precursor – the monoterpene iridoid loganin.

indole

The indole alkaloid structures typically involve multiple ring systems, often complex in character. They form the basis of several pharmaceutical drugs as well as some of our most potent hallucinogenic drugs, and include poisonous compounds such as strychnine. Of the indole alkaloids used in medicine and pharmacy, the majority are found in members of the family Apocynaceae (eg. *Rauvalfia, Vinca, Catharanthus* and *Alstonia* spp.). Other families in which they occur are the Loganaceae, Rubiaceae and Convovulaceae. Indole alkaloids are represented in the fungal kingdom by the ergot alkaloids and the *Psilocybin* mushrooms.

Reserpine

From the roots of *Rauvalfia serpentina* (Apocynaceae) along with related alkaloids **resicinnamine, deserpidine** and **ajmaline**. The main actions are hypotensive, sedative and tranquillizing. Ajmaline is of benefit for heart arrhythmias [Samuelsson 1992]

The primary actions of reserpine alkaloids are caused by inhibition of noradrenaline and depletion of amines in the central nervous system. While the hypotensive effects have a slow onset their duration is long, and the effective dose is sufficiently low to limit any side effects. The much higher doses

reserpine

required for tranquillisation which were once widely prescribed often resulted in depression and Parkinson Disease-like symptoms [Huang 1993]. As a result *Rauvalfia* has become a restricted herb in Australia and has no place in current herbal prescribing.

Alstonia

Alstonia constricta – Australian bitter bark

One of the best of the bitter tonics and febrifuge remedies from Australia. It is a uterine stimulant and should be avoided during pregnancy. It contains several antihypertensive alkaloids of the indole class – **alstonine, alstonidine, reserpine**.

Periwinkles

Catharanthus roseus – the rose or Madagascar periwinkle, a popular garden plant.

According to Samuelsson over 100 alkaloids have been isolated from this species, including some with hypoglycaemic properties. However the alkaloids occur naturally in very small quantities. **Vinblastine** and **vincristine** are anti-neoplastic, used as chemotherapy treatment in childhood leukemia and Hodgkin's Disease. They are dimeric indole alkaloids.

Vinca minor, V. major – the common periwinkles.

Numerous indole alkaloids including vincamine, majdine, majoridine.

Actions – antihaemorrhagic and astringent. *Vinca* spp. do not contain anti-neoplastic alkaloids found in *Catharanthus* [Wren 1988].

Ergot alkaloids

Ergot (*Claviceps purpurea*), a fungus with a sclerotina (fruiting body) that produces over 20 indole alkaloids. It grows on rye and other cereal plants. Ergotism is a toxic response to ingesting ergot contaminated grain, and manifests either as painful spasms of the limb muscles leading to epileptic-like convulsions ('St.Anthony's fire'), or as vomiting and diarrhoea leading to gangrene of the toes and fingers. Both syndromes can lead to fatalities.

Ergotamine constricts peripheral blood vessels and raises blood pressure. It is used in treatment of migranes. **Ergonavine** is oxytocic and vasoconstrictive. It is used as treatment or preventative for post-partum haemorrhage. **LSD (lysergic acid)** is a semi-synthetic derivative of ergot – similar compounds are found in *Ipomoea* spp. (morning glory).

Strychnine

From the seed of ***Strychnos nux-vomica*** (Loganiaceae), known as **nux vomica**.

Strychnine is a powerful stimulant to the central nervous system in low doses, but becomes deadly poisonous in larger doses (60-90mg), causing an exaggeration of reflexes and tonic convulsions [Tyler et al 1988]. **Brucine** is a related but less toxic alkaloid in **Strychnos** spp.

Yohimbine

Derived from the bark of *Aspidosperma quebracho* (Apocynaceae) and *Pausinystalia yohimba* (Rubiaceae). These plants are reputedly aphrodisiacs. Yohimbine blocks α-adrenergic transmissions but stimulates β-adrenergic sites [Robinson1981]. Yohimbine and strychnine are sometimes combined in pharmaceutical preparations for use as aphrodisiacs and nerve tonics.

STEROIDAL ALKALOIDS

Solanum alkaloids

Steroidal alkaloids are derived from triterpenoids, being distinguished from other compounds in that class by the presence of a nitrogen atom. The main sub-groups in this class are found in the Liliaceae families and Solanaceae. The Solanaceae group occur as glycosides – some are used as the basis for synthesis of steroid drugs. **Solasodine**, the glycoalkaloid used in production of contraceptives, is obtained from the Australian species *Solanum laciniatum* and *S. aviculare*, which are now cultivated in several countries for this purpose [Bradley et al 1978].

solasodine

In unripe potatoes (*Solanum tuberosum*) the glycoalklaoids α-**solanine** and α-**chaconine** are derived from the widely occurring aglycone solanidine. Glycoalkaloids are mainly concentrated in unripe fruits and green potatoes – they disappear in the ripening process. These alkaloids are toxic so that avoidance of sprouting potatoes and unripe tomatoes is recommended, particularly for pregnant women since there is evidence of teratogenicity. Supplementation with ascorbic acid has been shown to protect against toxicity from these alkaloids [Renwick 1986].

Antihepatoxic properties have been reported from two steroidal alkaloids in *Solanum capsicastrum* [Lin & Gan 1989].

Veratrum alkaloids

The *Veratrum* genus (Liliaceae) is also rich in steroidal alkaloids. They are derived from the white hellebore (*Veratrum album*) and the green hellebore (*V. viridis*). These alkaloids have steroid skeletons though some are highly oxygenated – the **protoveratrines A & B** contain up to nine oxygen atoms.

Veratrine alkaloids are cardiac depressants, and are used in medicine in the treatment of severe hypertension. Their action is mediated through inhibition of Na-K ATPase, the enzyme required for transport of mineral ions across cell membranes – an action they share with the cardiac glycosides [Robinson 1981]. Owing to their high toxicity the Veratrums are not used in herbal medicine.

ALKALOIDAL AMINES

These compounds are distinguished by the lack of heterocyclic nitrogen atoms – the N occurs in an amino side group. The precursors for alkaloidal amines are aromatic amino acids – phenylalanine, tyrosine and tryptophan.

Ephedrine

Ephedrine and **pseudoephedrine** are derived from the aerial parts of *Ephedra* spp. (Ephedraceae). The major source is the chinese species known as "ma huang" – *Ephedra sinica*, for centuries one of the most important medicines in the Chinese Materia Medica, whereas in the west its use is clouded in controversy and, in the case of Australia, restricted to registered practitioners. The two major alkaloids in *Ephedra* form the basis of several proprietary prescription medicines, as well as the illicit amphetamine drugs. Issues of safety concerns and legal status, as well as uses both ancient and modern, were recently reviewed in *Herbalgram* [Blumenthal & King 1995].

ephedrine

Ephedrine is structurally simple, its aromatic skeleton deriving from phenylalanine, while an extra methyl group (CH_3) derives from methionine [Samuelsson 1992]. The basic structure occurs in several isomeric forms, one of which (a stereoisomer) is **pseudoephedrine**.

Pharmacology of ephedrine

Ephedrine is a sympathomimetic or CNS stimulant. It is a potent stimulator of α, $\beta1$ and $\beta2$ adrenergic receptors. The effects include vasoconstriction, raised blood pressure and pulse, bronchodilation and diuresis. Ephedrine based drugs are used as nasal decongestants, bronchodilators and in anaphylactic shock. In excess they cause insomnia, tachycardia and dizziness.

In herbal medicine *Ephedra* is valued highly as a reliable treatment for asthma and allergic conditions of all types. Being both bronchodilatory and nasal decogestant (due to constriction of blood vessels) *Ephedra* is useful also in bronchitis, emphysema, rhinitis, as well as common colds and influenza. It is contraindicated for hypertension, angina pectoris, hyperthyroidism, during pregnancy and where MAO inhibitors are being used [Robson 1995].

Other alkaloidal amines

Colchicine – from *Colchicum autumnale* (Liliaceae) or autumn crocus.

Both the alkaloid and the herb are beneficial in the treatment of gout. Because of its ability to inhibit cell division, colchicine is used in plant breeding and genetic research.

Muscarine – from *Amanita muscaria* (Agaricaceae) or fly agaric mushroom.

Muscarine is a toxic constituent found in several mushroom species. It is a parasympathetic agent that binds to cholinergic receptors. Owing to its toxicity it is not used therapeutically.

Mescaline – from *Lophophora williamsii* (Cactaceae) – peyote, mescal buttons.

This plant is used as a hallucinogen by some North American Indians.

Psilocybin – from *Psilocybe* spp. and other hallucinogenic fungi.

Hordenine – from barley – *Hordeum vulgare* (Poaceae).

PURINE ALKALOIDS

Purines bases are a group of compounds found in plants and animals which includes the nucleic acids. Their biosynthesis is complex with numerous non-amino acid precursors [Samuelsson 1992]. **Xanthine**, an oxidised purine that occurs as a breakdown product of nucleic acid metabolism, is itself oxidised in the body to uric acid. Xanthine consists of two fused ring systems each containing two nitrogen atoms.

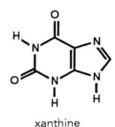

xanthine

Alkaloidal amines are methylated xanthines forming weak bases that are pharmacologically active. There are three methylxanthines, all are present in our most popular stimulant beverages – coffee and tea.

Caffeine

Found in a number of botanically unrelated species, including *Coffea arabica* (Rubiaceae), *Camellia sinensis* (Theaceae) or tea, *Cola nitida* (Sterculaciaceae) or kola nut and *Paullinia cupana* (Sapindaceae) or guarana. Caffeine is bound to chlorogenic acid in raw coffee beans, the roasting process liberating the caffeine and other compounds that contribute to the aroma of coffee [Samuelsson 1992].

caffeine

Caffeine is a CNS stimulant, enhancing alertness and overcoming fatigue, while high doses lead to insomnia and tremors. It also stimulates cardiac output and heart rate, and acts as a mild diuretic. Caffeine raises metabolism, influences blood sugar and is habit forming. It is sometimes used in formulations for treating migrane.

Theophylline

Structurally theophylline resembles caffeine, however it lacks the methyl group in the 5 carbon ring. It occurs naturally in the tea plant, however it is synthesized from caffeine for use in medicine. The effects on the CNS and cardio-vascular system are similar to those of caffeine though milder, while the diuretic activity is more pronounced. However the main use for theophylline is as a bronchial smoothe muscle relaxant for treatment of bronchial asthma and emphysema. It forms the basis of the drug aminophylline, used as a diuretic and asthma medicine.

Theobromine

Theobromine is found mainly in the cocoa plant, ie. *Theobroma cacao* (Sterculiaceae). It is isomeric with theophylline, however it lacks the potent CNS effects of the other two alkaloids in this class. It is used mainly as a diuretic and bronchial muscle relaxant.

References

Barnard, C. 1952. The Duboisias of Australia. *Economic Botany 6: 3-17.*

Blumenthal, M. & King, P. 1995. Ma Huang: ancient herb, modern medicine, regulatory dilemma. *Herbalgram* 34.

Bradley, P. (ed) 1992. *British Herbal Compendium Vol. 1.* Br. Herbal Med. Assn., U.K.

Bradley, V. Collins, D. Grabbe, P. et al. 1978. A survey of Australian Solanum plants for potential useful sources of solasidine. *Aust. J. Botany* 26:723-753

Bruneton, J. 1995. *Pharmacognosy Phytochemistry Medicinal Plants.* Lavoisier Pubs, Paris.

Chen, I. Wu, S. & Tsai, I. 1994. Chemical and bioactive constituents from *Xanthoxylum simulans. J. Natural Products* 57:1206-1211.

De Silva L.B. 1983. Medicinal plant research – retrospect and prospect. In R.S.Thakur et al (eds.) *Proceedings of International Symposium on Medicinal & Aromatic Plants.* Central Institute of Medicinal & Aromatic Plants, Lucknow.

Evans, W. 1990. Medicinal and poisonous plants of the Solanaceae. *Br J Phytotherapy* 1: 26-31

Grayer, R. & Harborne, J. A. 1994. Survey of antifungal compounds from higher plants. 1982-1983. *Phytochemistry* 37: 19-42

Harborne, J. & Baxter, H. 1993. *Phytochemical Dictionary.* Taylor & Francis, London.

Huang, K. 1993. *The Pharmacology of Chinese Herbs*. CRC Press, USA

Kapoor, L.D. 1995. *Opium Poppy. Botany, Chemistry, & Pharmacology*. Food Products Press, New York.

Koul, I. & Kapil, A. 1993. Evaluation of the liver protective potential of piperine, an active principle of black and long peppers. *Planta Medica* 59: 413-417.

Lin, C. & Gan, K. 1989. Antihepatotoxic principles of *Solanum capsicastrum*. *Planta Medica* 55: 48-50.

Marina Bettolo, G.B. 1986. Recent research on alkaloids active upon the central nervous system. In Barton & Ollis (eds.) *Advances in Medicinal Phytochemistry*. John Libbey & Co. London.

Martin, M. L. et al. 1993. Antispasmodic Activity of Benzylisoquinoline Alkaloids Analogous to Papaverine. *Planta Medica* 59: 63-67

Muller, K. & Ziereis, K. 1994. The Antipsoriatic Mahonia aquifolium and its active Constituents. *Planta Medica* 60: 421-424

Pizzorno, J. & Murray M. 1986. *Hydrastis, Berberis* and other berberine containing plants. In *Textbook of Natural Medicine*, John Bastyr College, Portland.

Plotkin, M. 1993. *Tales of a Shaman's Apprentice*. Penguin books, New York.

Renwick, J.H. 1986. Our ascorbate defense against the Solanaceae. In D'Arcy (ed) *Solanaceae. Biology and Systematics*. Columbia University Press, New York.

Robinson, T. 1981. *The Biochemistry of Alkaloids*. Springer-Verlag, Berlin.

Robinson, T. 1986. The Biochemical Pharmacology of Plant Alkaloids. In Cracker & Simon (eds.) *Herbs, Spices, & Medicinal Plants Vol. 1*. Oryx Press, Pheonix.

Robson, T. 1995. Ephedra sinica et spp. *Aust. J. Med. Herbalism* 7: 64-68.

Roddick, J. 1991. The importance of the Solanaceae in medicine and drug therapy. In Hawkes, et al. (eds) *Solanaceae III: Taxonomy, Chemistry, Evolution*. Royal Botanic Gardens, Kew UK.

Samuelsson, G. 1992. *Drugs of Natural Origin*. Swedish Pharmaceutical Press.

Schiff Jr. Paul L. 1991. Bisbenzylisoquinoline alkaloids. *Journal of Natural Products* 54: 645-749

Schmeller, T. et al. 1994. Binding of quinolizidine alkaloids to nicotinic and muscarinic acetycholine receptors. *Journal of Natural Products* 57: 1316-1319.

Sim, S. 1965. *Medicinal Plant Alkaloids.* University of Toronto Press, Canada.

Speisky, H. Squella, J. & Nunez-Vergara, L. 1991. Activity of boldine on rat ileum. *Planta Medica* 57: 519-522.

Toshiaki T. et al. 1993. Inhibitory effects of berberine-type alkaloids on elastase. *Planta Medica* 59: 200-202.

Tyler, V., Brady, L. & Robbers, J. 1988. *Pharmacognosy* 9th Edition. Lea & Febiger, Philadelphia.

Wagner, H. Bladt, S. & Zgainski, E. 1984. *Plant Drug Analysis.* Springer-Verlag. Berlin.

Watson, P. Luanratana, O & Griffen, W. 1983. The ethnopharmacology of pituri. *J. Ethnopharmacology* 8: 303-311.

Willard, T. 1992. *Textbook of Advanced Herbology.* Wild Rose College of Natural Healing. Alberta.

Wren, R.C. 1988. *Potter's New Cyclopaedia of Botanical Drugs and Preparations.* Saffron Walden , UK. Wu et al. 1989. Cytotoxicity of Isoquinoline Alkaloids and their N-Oxides. *Planta Medica* 55: 163-165

BIBLIOGRAPHY

Bradley, P. [ed] 1992. *British Herbal Compendium Vol.1.* British Herbal Medicine Assn., U.K.

Bruneton, J. 1995. *Pharmacognosy Phytochemistry Medicinal Plants.* Lavoisier Pubs, Pa.

Harborne, J. & Baxter, H. 1993. *Phytochemical Dictionary.* Taylor & Francis, London.

Huang, K. 1993. *The Pharmacology of Chinese Herbs.* CRC Press, USA

Mills, S. 1994. *The Essential Book of Herbal Medicine.* Viking Arkana, London.

Samuelsson, G. 1992. *Drugs of Natural Origin.* Swedish Pharmaceutical Press, Stockholm.

Schauenberg, P. & Paris, F. 1977. *Guide to Medicinal Plants.* Lutterworth Press.

Tyler, V. Brady, J. & Robbers, J. 1988. *Pharmacognosy 9th edition.* Lea & Febiger, Philadelphia.

Wagner, H. Sladt, S. & Zgainski, E. 1984. *Plant Drug Analysis.* Springer-Verlag. Berlin.

Wagner, H. & Wolff, P. [eds] *New Natural Products with Pharmacological, Biological or Therapeutic Activity.* Springer-Verlag, Berlin.

Willard, T. 1992. *Textbook of Advanced Herbology.* Wild Rose College of Natural Healing, Alberta.

Wren, R. 1988. *Potters New Cyclopaedia of Botanical Drugs and Preparations.* Saffron Walden, U.K.

INDEX

THE CONSTITUENTS OF MEDICINAL PLANTS

An Introduction to the Chemistry & Therapeutics of Herbal Medicines

The Author

The author, **Andrew Pengelly** B.A., N.D., D.B.M., D.Hom., was trained at the Southern Cross Herbal School and has been a practicing medical herbalist for 15 years. He has lectured widely in herbalism, botany and pharmacogosy.

Andrew is a founding editor of the *Australian Journal of Medical Herbalism* for which he continues to contribute regular articles. He served as an executive for the National Herbalist Association of Australia for ten years, and was recently made a fellow of that association. He is a graduate of the University of New England, with majors in plant biology and archaeology.

Andrew lives on a farm near Muswellbrook in the Hunter Valley where he grows medicinal herbs and organic vegetables.